CHROMATIN AND GENE FUNCTION

A Primer

CHROMATIN AND GENE FUNCTION

A Primer

John C Lucchesi
Emory University, USA

NEW JERSEY • LONDON • SINGAPORE • BEIJING • SHANGHAI • HONG KONG • TAIPEI • CHENNAI • TOKYO

Published by

World Scientific Publishing Co. Pte. Ltd.
5 Toh Tuck Link, Singapore 596224
USA office: 27 Warren Street, Suite 401-402, Hackensack, NJ 07601
UK office: 57 Shelton Street, Covent Garden, London WC2H 9HE

British Library Cataloguing-in-Publication Data
A catalogue record for this book is available from the British Library.

CHROMATIN AND GENE FUNCTION
A Primer

ISBN 978-981-125-843-5 (hardcover)
ISBN 978-981-125-893-0 (paperback)
ISBN 978-981-125-844-2 (ebook for institutions)
ISBN 978-981-125-845-9 (ebook for individuals)

For any available supplementary material, please visit
https://www.worldscientific.com/worldscibooks/10.1142/12894#t=suppl

I would like to dedicate this book to all of the members of my lab at the University of North Carolina in Chapel Hill and at Emory University in Atlanta, GA. They have been instrumental in shaping my scientific career and making research a life-long, exciting experience.

I am grateful to John Thomas Lucchesi and Home Team Graphics for creating the cover design and for producing original figures.

Preface

The purpose of this primer is to provide students, teachers as well as academic and industry researchers with a succinct account of the chemical and structural features of chromatin and the role that these features play in the maintenance and function of the genetic material. It is universally accepted that deoxyribonucleic acid (DNA) is the carrier of the genetic information that is transmitted from parents to their offspring. This information is responsible for the anatomy, physiology and behavior of all individuals throughout development and adult life. Not surprisingly, as it was formulated, this intellectually satisfying principle generated a large number of fundamental questions centered around the fact that, with the exception of mature ova and sperm, all of the cells of multicellular organisms have the same DNA molecules; yet, during development, groups of genetically identical cells differentiate to give rise to tissues and organs most of which contain a variety of distinct cell types that perform very different functions. How could cells extract specific sets of biological directives from a common genetic blueprint? Another longstanding unresolved question was whether and to what extent environmental circumstances could affect the stability of genetic inheritance. Answers to these questions have been provided by focusing on the longstanding observations that DNA is associated with a variety of different molecules — proteins and ribonucleic acids (RNAs) — forming a complex referred to as chromatin. The chemical modifications and selective interactions of these molecules with the DNA, as well as modifications of the DNA itself that do not alter the encoded information that it contains, provide the mechanisms responsible for cellular differentiation and for the environmental modulation of the genetic information. This level of regulation of the genetic blueprint is referred to as epigenetic regulation.

Although this primer's major focus is on the composition, structure and function of chromatin in mediating gene expression, an introductory section has been included that reviews general aspects of genetic inheritance. The purpose of this section is to establish the appropriate perspective for the presentation of the central theme. Boxes are inserted in some of the chapters for the purpose of expanding specific points without interrupting the flow of the narrative, and throughout the text, references are provided, generally to review papers, to enable a deeper exploration of the topics that are discussed.

Contents

CHAPTER 1

Introduction

The seminal discovery: Early in the course of human history, long before genetics was developed as a science, the inheritance of biological traits was used by nomadic tribes and by sedentary groups to select favorable characteristics in domesticated animals and cultivated plants. Over the centuries, the mechanisms responsible for heredity were the subject of various fanciful theories. Greek philosophers Hippocrates (famous for the Hippocratic oath, an oath of ethics traditionally taken in some form by medical school graduates) and Aristotle are credited with the first thoughts on reproduction. Hippocrates believed that all the organs of the parents' bodies produce "seeds" that are transmitted during mating; Aristotle believed that blood was responsible for determining an offspring's characteristics and that male semen was a purified form of blood.[1] Over the subsequent centuries, most hypotheses to explain heredity did not differ much from those of the Greek philosophers to whit Charles Darwin's belief that particles (gemmules) produced by each part of the body are included in the male and female gametes and are responsible for all the traits of the offspring. The foundation of modern genetics was established scientifically by Gregor Mendel, an Austrian monk residing in Brünn, Silesia (now known as Brno, Czech Republic). Starting in 1854, Mendel crossed varieties of garden peas that were visibly different in a particular characteristic such as flower color or seed color and was able to demonstrate that these different traits were transmitted independently of one another; he observed that one of the two

[1]This notion seems to have persisted to this day, witness the common expressions "blood relatives" and "blood lines".

forms of a parental trait that was present in a hybrid masked the presence of the other and called that form dominant while he termed the other recessive. In turn, the hybrid transmitted either the dominant or the recessive forms to each of its offspring. Mendel's experimental results [1,2], which he published in the journal of the *Natural Science Society* of Brünn, remained largely unnoticed and unappreciated for several decades. These results were eventually rediscovered by three botanists — Carl Correns, Erick von Tschermak and Hugo De Vries — who, independently and using different plant hybrids, had come to the same conclusions as Mendel. The transmitted factors that Mendel postulated were responsible for the various traits that he had studied were given the name of genes by Wilhelm Johannsen who also named the hereditary constitution of an individual its genotype and the resulting external characteristics its phenotype. Later, the different forms of a gene were referred to as alleles.

Genes and chromosomes: The next phase consisted of demonstrating that genes are present on chromosomes and that the behavior of chromosomes during cell division (mitosis) and especially during the formation of sex cells (meiosis) is concordant with the inheritance of genetic traits. Although the existence of chromosomes in various cell types had been noted by several scientists, their behavior during mitosis was first described by Walther Flemming who concluded that chromosomes split early in the process and that they are equally distributed to daughter cells (the chromosomes do not split; they replicate). Theodor Boveri recognized that the number of chromosomes is halved in sex cells (gametes) and is reconstituted after fertilization; he also noted that all the chromosomes need to be present to insure normal development [3]. Walter Sutton was able to follow individual chromosomes as they paired during the early stages of meiosis; he proposed that each of the chromosomes that had been inherited from one parent associated with a similar chromosome that had been inherited from the other parent and noted that each member of a pair has an equal chance to be included in a mature gamete [4]. These observations led Sutton to suggest that the behavior of chromosomes provided the physical basis for Mendel's rules of heredity. Several decades later, proof that genes are in fact located on chromosomes was offered by Thomas Hunt Morgan who showed that a recessive mutation that affected eye color was present on one of the sex chromosomes of fruit flies, the organism that he was using for genetic studies (sex chromosomes are chromosomes that are different in the two sexes). Other members of Morgan's research group demonstrated that different genes are located on the same chromosome [5]. During the formation of gametes, the alleles of genes present on one of the chromosomes of a pair could often be transmitted together to the offspring or they could be exchanged for the allele present on the other chromosome of the pair (the homologous chromosome). The recombination of alleles was shown to be the result of the physical exchange of segments between paired chromosomes by Barbara McClintock in plants and, independently, by Curt Stern in fruit flies [6,7]. The frequency of recombination was shown to depend on the physical distance between the genes in question.

The function of genes: The next major goal for genetic research was to determine the biochemical nature of genes and of their products. The first seminal observations in this area were made by a British physician, Alfred Garrod, who reported that a rare human condition consisting of the production of blackish urine (alkaptonuria) was hereditary and caused by a recessive "sport" (mutation). He further suggested that albinism (a failure in the synthesis of color pigment in the skin, hair and eyes) and cystinuria (a defect in kidney function resulting in the formation of aggregates or stones) may be additional examples of inherited errors in different metabolic pathways [8]. Two fundamental steps in understanding the relationship between genes and physiological pathways were the discovery of *deoxyribonucleic acid* (DNA) and its role in heredity, and that individual genes were involved in the production of proteins, many among which catalyzed metabolic reactions, i.e., were enzymes. Over the years, a number of individuals had suggested that genes were responsible, either directly or through the action of enzymes, for all of the chemical reactions that took place during the development and adult life of organisms. Working with the common mold Neurospora, George Beadle and Edward Tatum showed that X-ray exposure of spores could lead to very specific metabolic defects that would be inherited as if they were the result of mutations in single genes; their work introduced the concept of one gene–one enzyme–one biochemical reaction [9]. DNA was discovered long before its role in heredity was established. In 1869, Johann Mischer, a Swiss chemist, extracted a compound that he called nuclein, later identified as deoxyribonucleic acid. In 1928, an English scientist,

Frederick Griffith, reported that bacteria of a certain strain lethal to mice could transmit their virulence to an innocuous strain and transform it into a virulent strain. Several years later, the substance responsible for this transformation was shown to be DNA [10]. During the following decade, the chemical composition and physical structure of DNA molecules were elucidated, issuing in the modern phase of genetics.

DNA and the genetic code: The double helix model for DNA structure was devised by an American geneticist/biochemist, James Watson, and a British biophysicist, Francis Crick [11]. Their model was derived from X-ray diffraction[2] pictures of DNA by Maurice Wilkins, Rosalind Franklin and her graduate student Raymond Gosslin and, equally as important, from the older observation by Erwin Chargaff that the two types of nitrogenous bases that occur in DNA — purines and pyrimidines — are present in equal amounts. Purines (adenine [A] and guanine[G]) are larger molecules with two carbon–nitrogen rings; pyrimidines (thymine [T] and cytosine [C]) have a single ring. In the Watson and Crick model, the DNA molecule is a double helix made up of two strands of alternating sugar and phosphate molecules. On each strand, a nitrogenous base is attached to each sugar molecule and the building unit of the strand is the *nucleotide* (base–sugar–phosphate). A purine present on one strand is always paired with a pyrimidine on the other strand (Fig. 1). Each strand has a phosphate group ($-PO_4$) attached to the first sugar molecule (to the 5′ carbon) and a hydroxyl group (-OH) attached to the last sugar molecule (to the 3′ carbon) of the sugar–phosphate backbone (the sugar is *deoxyribose*). The two strands are parallel to each other but with opposite directionality (Fig. 1). One of the important features of the model was that it showed how DNA could be replicated: the strands could separate and each would direct the synthesis of a complementary strand. That DNA replication does, in fact, occur in this manner was demonstrated by one of the most elegant experiments of modern biology carried out by Matthew Meselson and Frank Stahl ([13]; see Box 1).

---

[2] X-ray diffraction is a technique based on the scattering of a beam of X rays by the atoms of a crystal. The scatter is recorded as a pattern of spots on a photographic plate that allows the measure of the angle and intensity of the diffracted X rays. These measurements reveal the position of the atoms within the crystal as well as their chemical bonds. A major challenge in the use of this technique is to obtain crystals of the molecules under study.

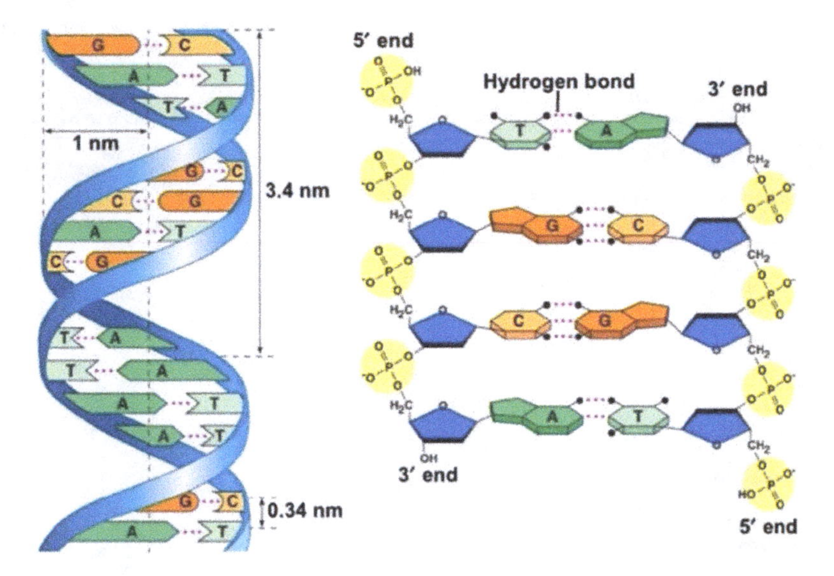

Fig. 1. DNA structure. Left: The molecule is made up of two chains of nucleotides. Right: Chemical representation of the nucleotide chains highlighting the sugar–phosphate backbone (yellow and blue molecules), the pairing of nitrogenous bases (A with T and G with C), and the reason for the opposite directionality of the two chains (from Kumar *et al.* [12]).

Box 1

The Meselson and Stahl experiment: Meselson and Stahl allowed a strain of bacteria to grow and reproduce for many generations in a medium that contained a form of nitrogen (N^{15}) that was heavier than normal nitrogen (N^{14}). When the DNA extracted from these bacteria was placed in a special solution and centrifuged at very high speed, it settled as a band that was faster and therefore was heavier than normal DNA. When the bacteria were raised for one generation on normal nitrogen, their DNA was lighter than the heavy DNA but heavier than normal DNA. A further generation in the presence of normal nitrogen led to an intermediate band and a second band similar to normal DNA (Fig. 2). These results showed that when DNA replicates, the two strands separate, and each strand directs the synthesis of a complementary strand. This mode of replication has been termed semi-conservative replication.

(*Continued*)

Box 1 (*Continued*)

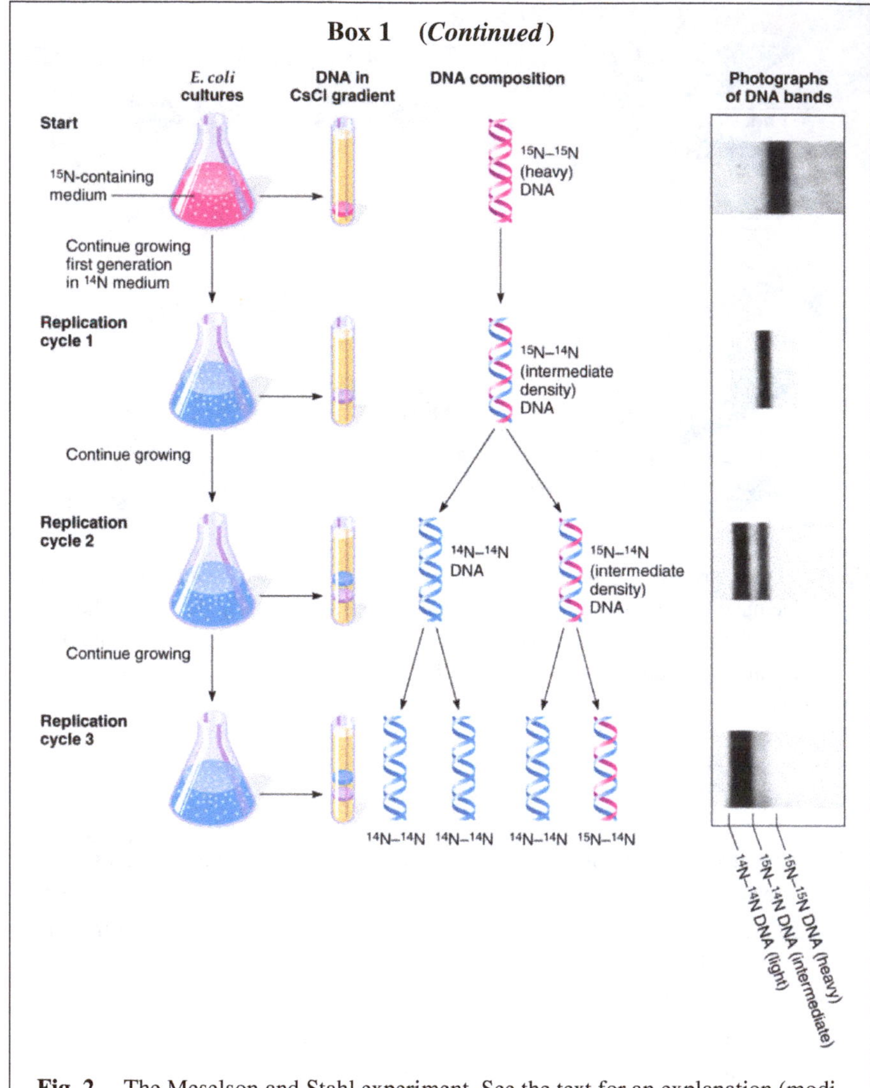

Fig. 2. The Meselson and Stahl experiment. See the text for an explanation (modified from *iGenetics*, 2nd ed.).

Although the direct relationship of genes and proteins was widely accepted, its nature was mysterious. A geneticist, Seymour Benzer, working with bacteriophages (bacterial viruses) had determined the position of different mutations that occurred within a particular gene; these observations led him to conclude that a gene was a sequence of nucleotides that could undergo a mutation affecting the protein that the gene specified [14]. In turn, proteins were known to consist of linear sequences of amino acids and a change in a single amino acid could lead to a malfunctioning, mutant protein [15]. Using this rather limited evidence as well as experimental data that they generated with bacteriophages, Francis Crick, Leslie Barnett, Sydney Brenner and Richard Watts-Tobin collaborated in an amazing exercise of deductive logic to propose the genetic code. They postulated that a sequence of three nitrogenous bases along the DNA region of a gene corresponded to a given amino acid in the protein specified by the gene. Because there are four different nitrogenous bases and therefore 64 possible combinations of three consecutive bases and there are only 20 amino acids, they suggested that more than one base triplet could call for a specific amino acid [16]. Subsequent work would show that the genetic code is the same in all organisms (see Box 2).

The flow of genetic information: It was still not clear how the genetic information encoded in DNA that is found in the nucleus of cells is transferred to the cytoplasm where protein synthesis occurs. The answer to this puzzle was provided by the discovery of *messenger ribonucleic acid* (mRNA). The first inkling of the existence of this RNA occurred to two virologists, Elliott Volkin and Lazarus Astrachan, who noted that two types of RNAs were produced in bacteria following bacteriophage infection, a stable type and a type that turned over rapidly. This latter RNA was more similar in nitrogenous base composition to the phage than to the bacterial DNA and they suggested that it might be responsible for the synthesis of phage-specific proteins [17]. The existence of mRNA [18]

Box 2

The genetic code: This code consists of sequences of three consecutive DNA nucleotides (triplets or codons). Since there are four different nucleotides, the total of possible codons is 64. Given that there are 20 amino acids, it is clear that each amino acid is specified by more than one codon (Fig. 3).

(*Continued*)

Box 2 (*Continued*)

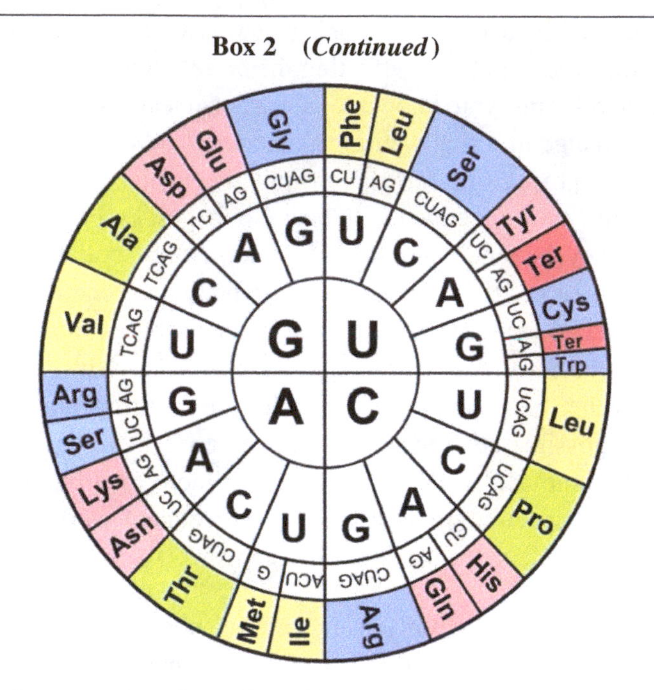

Fig. 3. Representation of the 64 possible codons with the amino acids that they specify. For a particular amino acid, listed by its abbreviation in the outer circle, the specific codons are determined by selecting a letter from the inner circle, followed by a letter from the second circle and one of the letters from the third circle. For example, two codons represent the code for glutamine (GLU): GAA and GAG. Note that three codons (UAA, UAG and UGA) represent termination codons, labeled Ter in the outer circle; the presence of any one of these codons in an mRNA molecule stops the synthesis of the protein (from Saier *et al.*, 2019[a]).

[a]Saier Jr, M.H., Understanding the genetic code. *Journal of Bacteriology*, 2019. **201**(15): 1–12.

and its role in protein synthesis [19] were established in the same year that the genetic code was proposed, introducing the central tenet of genetics: DNA information is copied unto mRNA that, in turn, is used as a template for the synthesis of proteins. Another major advance that occurred that year was the evidence obtained in bacteria by two French scientists, François Jacob and Jacques Monod, that the genes that specify the synthesis of proteins are under the control of other genes that encode regulatory molecules. These molecules can be activated or inactivated by associating with specific metabolites thereby regulating the synthesis by the gene of a messenger that is targeted to the site of protein synthesis, the ribosomes [20]. These conclusions established the basic model for all gene regulation.

Molecular cloning and genome sequencing: The development of molecular cloning techniques — specifically DNA cloning — enabled the next major advances in the study of genetics. The cornerstone of this experimental approach was the discovery by Hamilton Smith of bacterial enzymes, termed restriction enzymes, that cleave foreign DNA at specific nucleotide sequences [21]. If the DNA fragments generated by digestion with these enzymes contained a protein binding site, they could be isolated as protein–DNA complexes; fragments that contained particular genes could be identified by their association with their mRNA transcripts. Specific fragments could be inserted into the DNA of bacteriophages or plasmids (self-replicating, small DNA circles that are present in bacteria) and, following extensive bacterial replication, could be reisolated in large quantities. The massive amplification of specific DNA regions could also be obtained by a technique termed the polymerase chain reaction (PCR) that uses temperature cycles to unwind the DNA molecules and a heat-resistant enzyme that replicates the DNA strands [22]. The availability of amplified DNA molecules allowed the development of methods to determine the actual sequence of bases along a DNA region. These types of data were crucial in order to deduce the coded information in the region and to determine the occurrence and nature of mutations. The original sequencing method devised by Allan Maxam and Walter Gilbert relied on the chemical digestion of the DNA at base-specific sites [23] and was soon supplanted by the more efficient method of Frederick Sanger [24]. This method used the replication of the DNA section under study in the presence of modified nitrogenous bases that would stop the replication at the point where they were incorporated. Improvements in the Sanger method and its automation led to the sequencing of several organisms with increasing genetic complexity culminating in 2001 in the sequencing of the human genome [25,26]. Since that time, new techniques and complete automation have increased the speed and lowered the cost of DNA sequencing to the point where individual genes and entire genomes can be sequenced routinely.

The genetics of populations: While investigations at the cellular and molecular levels were being pursued, the study of the genetic composition of natural populations of organisms was proceeding in parallel. The impetus behind this area of genetics was the need to develop a mechanistic basis for Charles Darwin's earlier observations on the role of natural selection in the process of evolution. In those early days, the challenge was to explain the observations of continuous phenotypic variation in populations (such as the height of individuals) on the basis of discrete hereditary particles (genes). As Mendel had shown, these genes could lead through mutations to discontinuous traits between parents and offspring. The first explanation of this puzzle was provided by a British statistician/geneticist, Ronald A. Fisher, who proposed that continuous traits are the result of many genes, each contributing slightly to the trait [27]. During the following years, Fisher's work was expanded and complemented by John B. S. Haldane and Sewall Wright [28,29]. The work of these individuals and others, notably Theodosius Dobzhansky [30], led to the formulation of basic factors that can influence the genetic constitution of populations over time and, therefore, affect evolution: population size, mutation, genetic drift (fluctuations in the transfer of alleles from one generation to the next), natural selection, environmental diversity, migration and non-random mating patterns. This fusion of Darwinian observations with Mendelian genetics has been termed the "modern synthesis".

Not surprisingly, the field of population genetics has undergone major changes, some driven by new technologies and others by the realization that additional biological phenomena may impact the evolution process. The first molecular technique used to study variations of gene products (allelic versions of different enzymes) in natural populations was gel electrophoresis[3] [31]. Following the inception of cloning and DNA sequencing

---

[3]Gel electrophoresis is a technique that is used to separate a mixture of molecules (nucleic acids or proteins) or their fragments on the basis of their electrical charge and size.

techniques, the study of individual loci in populations has been extended to the analysis of genome-wide DNA variation [32]. New concepts that have been taken into consideration in the study of populations and in the development of modern evolution theory are cooperation and altruism, the notion of evolvability, i.e., the natural inclination of living organisms to evolve, and the feedback that modifications of the environment caused by organisms' adaptations have on the continued evolution on these adaptations [33,34]. Among the recently expanded areas of biology that influence the study of evolutionary processes are the developmental modalities and developmental plasticity of organisms, as well as the epigenetic regulation of gene function.

Epigenetics and the role of chromatin in gene function: The level of understanding of the basic mechanisms of heredity and of gene function represented amazing achievements in biology. Nevertheless, how the genetic blueprint is used to achieve the development of organisms and their physiology and behavior during adult life was still largely unexplained. The problem, of course, was that although all cells of an embryo contain identical genetic information, as development proceeds, cells and tissues differentiate. How are the genes that are needed to generate specific cell types activated at particular times and in particular regions of an embryo? The first insight into the mystery of differential gene activity was provided by two biochemists, Vincent Allfrey and Alfred Mirsky, who showed that some of the basic proteins known as histones that are associated with DNA were modified in metabolically active cells: histones needed to be acetylated to allow the DNA to serve as a template for RNA synthesis [35]. A few years later, Arthur Riggs and Robin Holliday independently proposed that DNA methylation, a modification common in bacteria and other organisms, could be responsible for repressing gene activity [36,37]. DNA methylation does not alter the genetic code of the DNA but interferes with the binding of activating regulatory proteins. All of these observations highlighted the importance of the fact that DNA does not exist in isolation in the nuclei of cells; rather it is associated with different histones and other proteins, as well as different types of regulatory RNA molecules in a complex referred to as chromatin. The first evidence that active and inactive genes are structurally different was obtained by Harold Weintraub and Mark Groudine who demonstrated that the DNA of active genes in purified nuclei can be more easily enzymatically digested [38]. Separate discoveries launched the modern field of epigenetic regulation of gene expression.

The first multiprotein complex that appeared to regulate the transcription of many genes was isolated and characterized by Craig Peterson and Ira Herskowitz [39]. This complex acted by facilitating the association of activators with the DNA sequences of particular genes. A few years later, James Brownell and C. David Allis discovered the first enzyme responsible for the acetylation of histones in chromatin. They showed that this histone acetyltransferase was the ortholog[4] of a factor long known to be necessary for the activation of a large group of genes in yeast [40,41]. Numerous additional modifications of histones and the enzymes responsible for inducing or eliminating these modifications have been identified. In the same vein, numerous multiprotein complexes that remodel the conformational state of chromatin have been characterized. Newly developed bioinformatics approaches[5] have demonstrated the extensive structural and functional conservation of all chromatin modifications and of the responsible factors and have confirmed some general rules for gene function: initiating factors recognize and bind to specific DNA sequences and recruit the complexes that modify chromatin and mediate gene activation or repression. The differential gene activity that is responsible for cell, tissue and organ differentiation during development is triggered by the nonuniform distribution of initiating factors during egg formation.

The chromatin modifications just discussed — DNA methylation, chemical changes in proteins or architectural reorganization of chromatin components — do not alter the genetic code, in other words, they are not mutations. They are said to represent epigenetic modifications involved in the epigenetic regulation of gene function. The term was coined by British

[4] Orthologs are genes that are derived by evolution from a common ancestral gene and that have retained the same function.

[5] Bioinformatics is an interdisciplinary field that consists of developing mathematical and statistical formulas as well as computer programs for the interpretation of large biological data sets such as the DNA sequence of whole genomes and global epigenetic patterns across whole chromosomes or entire genomes.

developmental biologist Conrad Waddington to describe all of the external influences that collaborate with the genetic blueprint to achieve animal development [42]. It is now clear that these influences are exercised through chromatin modifications. Waddington further proposed that developmental changes induced by environmental conditions could, in certain circumstances, become heritable. This contention is currently being widely investigated given its importance in human development and health as well as its potential impact on evolution theory.

References

[1] Mendel, G., *Experiments in Plant Hybridization*, 1865. www.esp.org/foundation/genetics/classical/gm-65.pdf.

[2] Abbott, S., and D.J. Fairbanks, Experiments on plant hybrids by Gregor Mendel. *Genetics*, 2016. **204**(2): 407–422.

[3] Baltzer, F., Theodor Boveri. *Science*, 1964. **144**(3620): 809–815.

[4] Sutton, W.S., On the morphology of the chromosome group in *Brachystola magna. Biological Bulletin*, 1902. **4**(1): 24–39.

[5] Morgan, T.H., Random segregation versus coupling in Mendelian inheritance. *Science*, 1911. **34**(873): 384.

[6] Creighton, H.B., and B. McClintock, A correlation of cytological and genetical crossing-over in Zea mays. *Proceedings of the National Academy of Sciences, U.S.*, 1931. **17**(8): 492–497.

[7] Stern, C., Factorenaustausch und austausch von chromosomenstucken. *Forschungen fortschr*, 1931. **7**: 447–448.

[8] Garrod, A., The incidence of alkaptonuria: A study in chemical individuality. *Lancet*, 1902. **2**: 1616–1620.

[9] Beadle, G.W., and E.L. Tatum, Genetic control of biochemical reactions in Neurospora. *Proceedings of the National Academy of Sciences, U.S.*, 1941. **27**(11): 499–506.

[10] Avery, O.T., *et al.*, Studies on the chemical nature of the substance inducing transformation of Pneumococcal types. *Journal of Experimental Medicine*, 1944. **79**(2): 137–158.

[11] Watson, J.D., and F.H.C. Crick, Molecular structure of nucleic acids. *Nature*, 1953. **171**(4356): 737–738.

[12] Kumar, V., *et al.*, DNA nanotechnology for cancer therapy. *Theranostics*, 2016. **6**(5): 710–725.

[13] Meselson, M., and F.W. Stahl, The replication of DNA in *Escherichia coli*. *Proceedings of the National Academy of Sciences, U.S.*, 1958. **44**(7): 671–682.

[14] Benzer, S., On the topology of the genetic fine structure. *Proceedings of the National Academy of Sciences, U.S.*, 1959. **45**: 1607–1620.

[15] Ingram, V.M., Gene mutations in human haemoglobin: The chemical difference between normal and sickle cell haemoglobin. *Nature*, 1957. **180**(4581): 326–328.

[16] Frick, F.H.C., *et al.*, General nature of the genetic code for proteins. *Nature*, 1961. **192**(4809): 1227–1232.

[17] Volkin, E., L. Astrachan, and J.L. Countryman, Metabolism of RNA phosphorus in *Escherichia coli* infected with bacteriophage T7. *Virology*, 1958. **6**: 545–555.

[18] Brenner, S., F. Jacob, and M. Meselson, An unstable intermediate carrying information from genes to ribosomes for protein synthesis. *Nature*, 1961. **190**(4776): 576–581.

[19] Nirenberg, M.W., and J.H. Matthaei, The dependence of cell-free protein synthesis in *E. coli* upon naturally occurring or synthetic polyribonucleotides. *Proceedings of the National Academy of Sciences, U.S.*, 1961. **47**(10): 1588–1602.

[20] Jacob, F., and J. Monod, Genetic regulatory mechanisms in the synthesis of proteins. *Journal of Molecular Biology*, 1961. **3**: 318–356.

[21] Nathans, D. and H.O. Smith, Restriction endonucleases in the analysis and restructuring DNA molecules. *Annual Reviews of Biochemistry*, 1975. **44**: 273–293.

[22] Saiki, R.K., *et al.*, Enzymatic amplification of beta-globin genomic sequences and restriction site analysis for diagnosis of sickle cell anemia. *Science*, 1985. **230**(4732): 1350–1354.

[23] Maxam, A.M., and W. Gilbert, A new method for sequencing DNA. *Proceedings of the National Academy of Sciences, U.S.*, 1977. **74**(2): 560–564.

[24] Sanger, F., S. Nicklen, and A.R. Coulson, DNA sequencing with chain-terminating inhibitors. *Proceedings of the National Academy of Sciences, U.S.*, 1977. **74**(12): 5463–5467.

[25] International Human Genome Sequencing Consortium, Initial sequencing and analysis of the human genome. *Nature*, 2001. **409**(6822): 860–921.

[26] Venter, J.C., *et al.*, The sequence of the human genome. *Science*, 2001. **291**(5507): 1304–1351.

[27] Fisher, R.A., The correlation between relatives on the supposition of Mendelian inheritance. *Transactions of the Royal Society of Edinburgh*, 1919. **52**(2): 399–433.

[28] Haldane, J.B.S., A mathematical theory of natural and artificial selection-I. *Transactions of the Cambridge Philosophical Society*, 1924. **23**: 19–41.

[29] Wright, S., Evolution in Mendelian populations. *Genetics*, 1931. **16**: 97–159.

[30] Dobzhansky, Th., *Genetics and the Origin of Species*, 1937. Columbia University Press, New York (2nd ed., 1941; 3rd ed., 1951).

[31] Smithies, O., Zone electrophoresis in starch gels: Group variations in the serum proteins of human adults. *Biochemistry*, 1955. **61**(4): 629–641.

[32] Huang, W., *et al.*, Natural variation in genome architecture among 205 *Drosophila melanogaster* genetic reference panel lines. *Genome Research*, 2014. **24**(7): 1193–1208.

[33] Pigliucci, M., Do we need an extended evolutionary synthesis? *Evolution*, 2007. **61**(12): 2743–2749.

[34] Laland, K.N., *et al.*, The extended evolutionary synthesis: Its structure, assumptions and predictions. *Proceedings of the Royal Society B: Biological Sciences*, 2015. **282**(1813): 1–14.

[35] Pogo, B.G.T., V.G. Alfrey, and A.E. Mirsky, RNA synthesis and histone acetylation during the course of gene activation in lymphocytes. *Proceedings of the National Academy of Sciences, U.S.*, 1966. **55**(4): 805–812.

[36] Riggs, A.D., X inactivation, differentiation, and DNA methylation. *Cytogenetics and Cell Genetics*, 1975. **14**(1): 404–420.

[37] Holliday, R., and J.E. Pugh, DNA modification mechanisms and gene activity during development. *Science*, 1975. **187**(4173): 226–232.

[38] Weintraub, H., and M. Groudine, Chromosomal subunits in active genes have an altered conformation. *Science*, 1976. **193**(4256): 848–856.

[39] Peterson, C.L., and I. Herskowitz, Characterization of the yeast SW1, SWI2 and SWI3 genes which encode a global activator of transcription. *Cell*, 1992. **68**(3): 573–583.

[40] Brownell, J.E., and C.D. Allis, An activity gel assay detects a single catalytically active histone acetyltransferase subunit in *Tetrahymena macronuclei. Proceedings of the National Academy of Sciences, U.S.*, **92**(14): 6364–6368.

[41] Brownell, J.E., *et al.*, Tetrahymena histone acetyltransferase A: A homolog to yeast Gcn5p linking histone acetylation to gene activation. *Cell*, 1996. **84**(6): 843–851.

[42] Waddington, C.H., Selection of the genetic basis for an acquired character. *Nature*, 1952. **169**(4294): 278.

CHAPTER 2

The Genetic Basis of Biological Heredity

A. Classical Transmission Genetics

Genes are sequences of DNA that contain the code for molecules (proteins or regulatory RNAs) that perform a biochemical function. The sequence of a gene can change (can mutate) and each new version of the sequence represents an *allele* of the gene. In *diploid* organisms (organisms that inherit one set of genes from each parent), genes are represented by two alleles. If these alleles are the same, the individual is *homozygous* and if they are different, the individual is *heterozygous* for the gene. In heterozygous individuals, one allele (*dominant* allele) can hide the presence of the other (*recessive* allele). When a heterozygous individual reproduces, there is an equal chance that it will include either of the two alleles in his or her sex cells (gametes: sperm and eggs) and, therefore, that it will transmit one or the other allele to an offspring.

Genes are present on *chromosomes*. Each chromosome contains a single molecule of DNA that is associated with *histones* (highly basic) and numerous non-histone proteins and regulatory RNA molecules (see Chapter 3) forming the molecular complex known as *chromatin*. The alleles of genes present on different chromosomes are transmitted independently to the next generation (Fig. 1). The alleles of genes that are present on the same chromosome (i.e., on the same DNA molecule) can

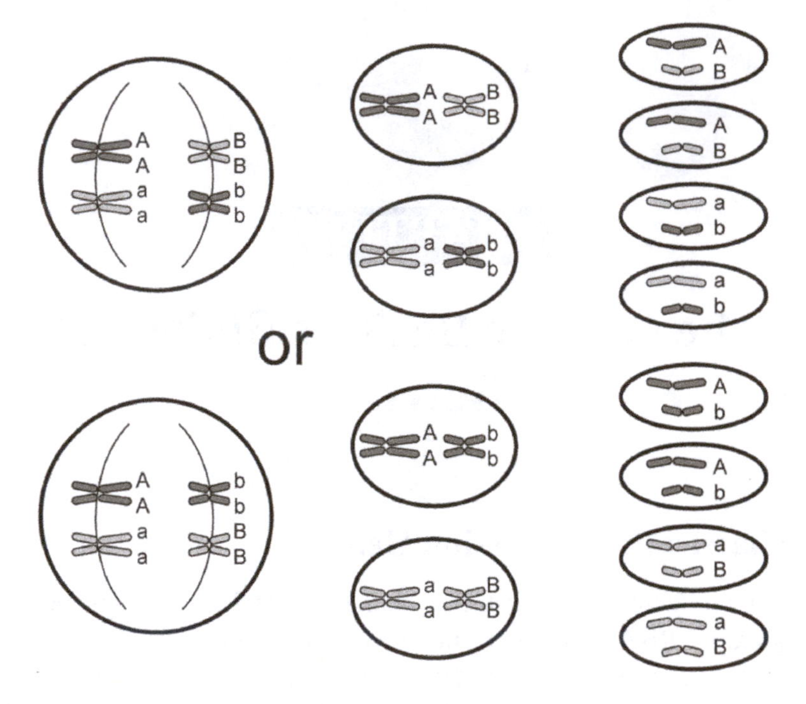

Fig. 1. Independent assortment of alleles of two genes during meiosis and the formation of gametes. When sex cells (spermatogonia in males and oogonia in females) enter meiosis, chromosomes have replicated and homologous chromosomes are paired. During the first meiotic division, each replicated chromosome has an equal chance of being associated with any other replicated chromosome. This insures that each of the two alleles of one gene can be included in the games and transmitted to progeny with either of the alleles of the gene on the other chromosome.

be passed on together to offspring; this occurs most of the time if the genes are close to each other on the chromosome but less frequently if they are more distant. The farther apart two genes are, the more frequently the allele of the gene present on one chromosome will be inherited with the allele of the other gene that is present on the *homologous chromosome* (the similar chromosome inherited from the other parent). This recombination of alleles is achieved through the process of *crossing over*, a physical exchange of regions between homologous chromosomes (Fig. 2).

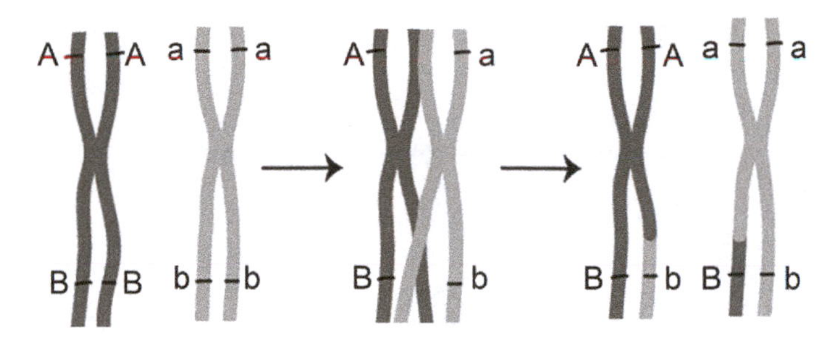

Fig. 2. During meiosis, when the replicated homologous chromosomes are paired, a physical exchange of chromosome sections can occur between them, leading to a new combination of alleles of the genes present on the chromosomes, that can be passed on to offspring.

Organisms are divided into two major groups depending on whether their genetic material is present free in the cell or is enclosed within a nucleus. Members of the former group, known as *prokaryotes*, are single-celled organisms, such as bacteria. Members of the latter group, the *eukaryotes*, can be single-celled, such as protozoa, yeasts and some algae, or multicellular. If eukaryotes receive one set of chromosomes from each parent, they are said to be diploid; if they receive multiple sets, they are said to be *polyploid*. In sexually reproducing eukaryotes, the sets of chromosomes received from each parent are identical (*autosomes*) except for one pair, the *sex chromosomes*. Different configurations of sex chromosomes exist in nature. In many mammals including humans, some insects, such as fruit flies, some reptiles and fishes, males have two sex chromosomes that are different and are called the X and Y chromosomes; females have two X chromosomes. In birds and some species of fishes and insects, it is the females that have the different sex chromosomes (Z and W), while in spiders and some insects, a single sex chromosome is present in one sex, while the other sex has two copies. In addition to the genes involved in the determination of sex, there are usually many genes present on the sex chromosomes that specify general biological functions. Since the sex chromosomes are different in the two sexes, of necessity, the inheritance of the alleles representing these genes will be associated with the inheritance of sex (Fig. 3).

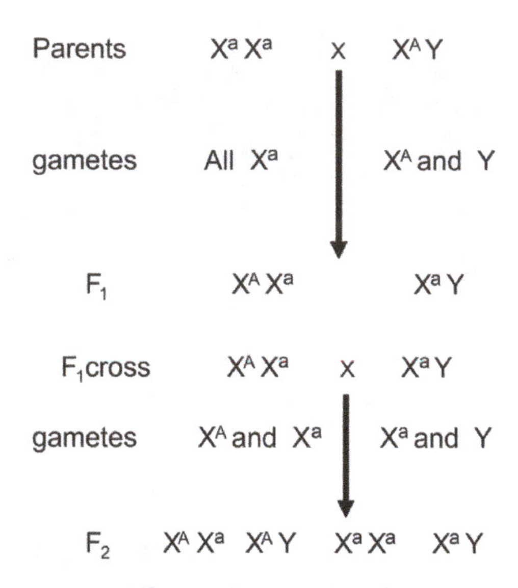

Fig. 3. An example of sex-linked inheritance. Females are homozygous for the recessive allele of a gene that is present on the X but is absent on the Y chromosomes. Males carry a dominant allele of the gene; F1 and F2 are the symbols indicating the first and second filial generations, respectively. The phenotype of the parents of the original cross (recessive in females, dominant in males) is reversed in the F1 generation. In the F2 progeny, half of each sex has the dominant phenotype and half has the recessive phenotype.

B. Cytogenetics

Cytogenetics started as the branch of genetics that studies the number and structural variations of chromosomes. Following the advent of molecular techniques, cytogenetic studies encompassed the localization of specific DNA sequences on chromosomes, as well as the high-resolution detection of copy number changes of DNA sequences in entire genomes. The cells of eukaryotic organisms always contain the precise number of chromosomes that are characteristic of the species to which they belong. In diploid organisms, the chromosomes consist of two *haploid* sets inherited from the organism's parents. Changes in the number of chromosomes, or duplication, deletion and rearrangement of the genetic material that they contain, can occur. These variations, referred to as chromosome

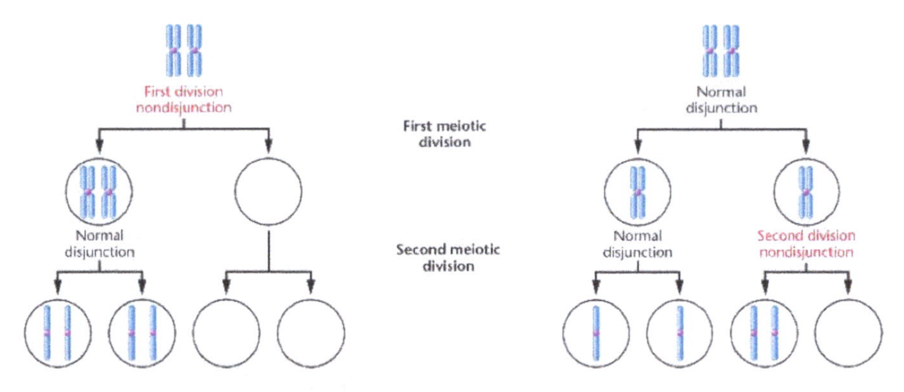

Fig. 4. Non-disjunction. Meiosis consists of two divisions; during the first division, MI, the replicated homologous chromosomes separate from each other; during the second division, MII, the two copies of each chromosome (chromatids) separate and are included into different gametes. Non-disjunction during MI (left panel): the two homologous chromosomes fail to separate. Non-disjunction during MII (right panel): the two chromatids of one chromosome fail to separate. Fertilization of a non-disjunction gamete with two copies of a given chromosome will give rise to a trisomic progeny individual; if the non-disjunction gamete has no copy of a chromosome, fertilization will lead to a monosomic individual. (From Essentials of Genetics, Sixth Edition).

mutations, usually have significant consequences ranging from the production of inviable gametes to developmental abnormalities of the offspring. The most common cause of chromosome number variation is non-disjunction during meiosis (Fig. 4). Other chromosome mutations occur when chromosomes are accidentally broken by the exposure of cells to certain chemicals or certain types of radiation. Such breaks may result in the loss (deletion) or the presence of an additional chromosomal region (duplication). Breaks may also lead to the exchange of regions between non-homologous chromosomes (translocations).

The localization of specific DNA sequences along the genome requires the use of two molecular techniques: DNA *cloning* and *in situ hybridization.* The linchpin of DNA cloning is the use of *restriction enzymes* that recognize a short sequence of nucleotides on the DNA molecule and make a staggered cut, i.e., they leave a single-stranded overhang at the site of the cut (Fig. 5). These overhangs allow the insertion of a

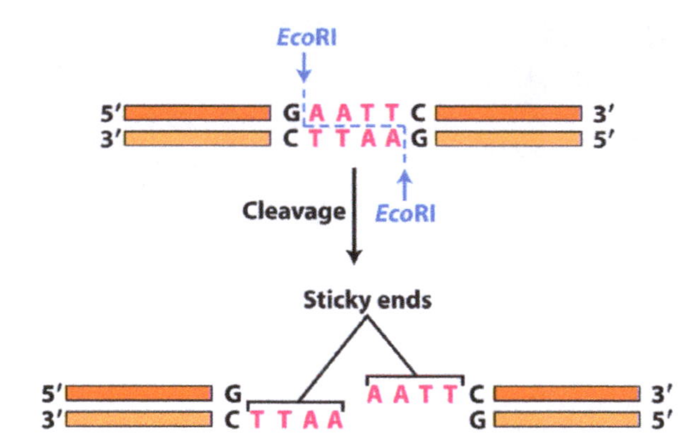

Fig. 5. Example of the DNA cut effected by a restriction enzyme (EcoRI). The recognition site of the enzyme is a six-nucleotide sequence. The cuts on the two DNA strands are staggered so that two single-stranded ends ("sticky ends") are left on the two cut molecules (From Molecular Cell Biology, 6th Edition).

DNA fragment to be cloned into a bacterial plasmid that is cut with the same enzyme (Fig. 6).

In situ hybridization consists of using a labeled DNA or RNA sequence (probe) to identify the location of the naturally occurring sequence in a biological sample, such as a chromosome, whole cell or tissue sample. Initially, the labels used were radioactive nucleotides, but more recently, the labels are chemical modifications that either fluoresce or that can be recognized by fluorescing antibodies. The DNA of the probe and target must be denatured (the complementary strands must be separated) to allow annealing of their complementary sequences.

The principle of *in situ* hybridization has been used for more fine-grained genome analyses [1]. One example is the high-resolution detection of copy number changes of DNA sequences in entire genomes. The DNA under investigation, labeled with a fluorescent dye of a specific color, is compared to a reference DNA, labeled with a dye of a different color, by hybridizing both to a microarray — a collection

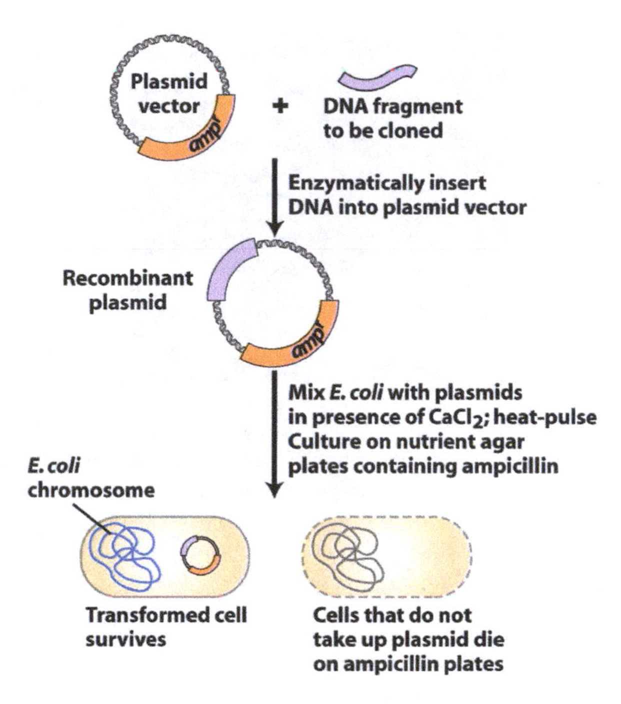

Fig. 6. Cloning of a fragment. The fragment was produced by cutting a DNA molecule with a restriction enzyme. The plasmid used for the cloning is cut with the same restriction enzyme so that both fragment and plasmid have complementary sticky ends. The transformed bacterial cells are allowed to reproduce extensively, the plasmids that they contain are isolated and the cloned DNA fragment is obtained by cutting the plasmid with the original restriction enzyme. (From Molecular Cell Biology, 6th Edition).

of known DNA sequences affixed as separate spots, in ordered fashion, on a glass slide. A difference in the level of hybridization of the two DNA samples at a particular spot can detect the presence of deletions or duplications in the test DNA (Fig. 7). Another example of the use of *in situ* hybridization is the detection of *single nucleotide polymorphisms* (SNPs) between members of a species. In this case, the microarray consists of total DNA or of DNA samples that contain specific sets of known SNPs.

Fig. 7. The diagram represents the results of a hybridization experiment on a DNA microarray. The reference and test DNA samples were labeled with red and green fluorescing dyes, respectively. A yellow spot represents a sequence present in both DNA samples; a red spot represents a sequence that was hybridized by the reference DNA and is missing in the test DNA; a green spot represents a sequence that was present in excess in the test DNA. (From Molecular Cell Biology, 6th Edition).

C. Molecular Genetics

Molecular genetics is the study of the structure and activity of genes, and of the mechanisms by which they mediate cellular function and organismal development, at the molecular level. A long chain of discoveries, both factual and technical, have led to the development of this approach to the study of biological inheritance. The following chapter (Chapter 3) reviews the molecular mechanism of *transcription* — the copying of a gene's DNA sequence to make an RNA molecule — and the remainder of this book will be a comprehensive outline of the molecular composition of chromatin and its role in regulating gene function. Therefore, this section will be limited to an overview of the major technical advances that have underpinned the development of molecular genetics.

High-throughput sequencing [2] relies on a number of very different techniques that share the following general steps: (i) the DNA to be sequenced is sheared into random fragments and adaptors are attached to the fragments that allow them to be bound to some surface, (ii) each DNA fragment is replicated numerous times leading to clusters of identical fragments (referred to as clonal amplification) and (iii) the fragments are sequenced by synthesis or by hybridization. One version of the former approach consists of adding labeled nucleotides one at a time, recording the incorporation and removing the label before adding the next nucleotide. Sequencing by hybridization involves the successive addition of labeled single or two-base labeled probes; as in the previous technique, the labels are removed after hybridization. RNA samples can be sequenced in a similar fashion by first converting them into copy DNA (cDNA). Processing and interpreting the massive amounts of data produced by these techniques have required the collaboration of mathematicians and statisticians and the development of the field of bioinformatics.

In parallel with the generation of nucleic acid sequencing methods, new technologies for the detection, identification and quantification of proteins have been developed. At the center of this area of research is *mass spectrometry*, an analytical tool that relies on measuring the mass and the charge of one or more proteins in a sample and identifying them on the basis of the ratio of these values [3]. In preparation for mass spectrometry, a number of steps that involve different types of separation techniques based on the relative charge of different molecules can be taken to "clean up" the sample. Mass spectrometers contain three components: a source of ions with which to bombard the sample and convert it into a gas, a mass analyzer that separates the ions of the gas phase on the basis of their mass and charge, and a detection system that calculates the relative abundance of the different ions.

Important advances have occurred in microscopy technology, based on new instrumentation. For example, *photo-activated localization microscopy* (PALM) and *stochastic optical reconstruction microscopy* (STORM) are imaging methods based on the use of fluorescent signals. In classical fluorescent microscopy, fluorescent stains or fluorescence-labeled antibodies are used to bind to specific molecules in cells. Upon exposure to the appropriate exciting light (usually ultraviolet light), the molecules fluoresce and become visible through a microscope. The problem is that the proximity of countless molecules fluorescing simultaneously reduces the resolution with which they can be detected and analyzed. In PALM or STORM, by using very short bursts of exciting light, very few molecules are allowed to fluoresce and be photographed at any one time, increasing the resolution by orders of magnitude.

A burgeoning area of study in molecular genetics is the study of the physical characteristics of molecules and complexes that constitute chromatin. For example, changes in the conformation of chromatin fibers — DNA molecules and their associated histone and non-histone proteins — can be measured by tethering them between a glass surface and a microbead and then moving the microbead. This can be accomplished either by placing it in a magnetic field or by using the force of a magnetic field generated by a laser beam (Fig. 8).

RNA interference is an example of a number of ingenious biological processes that have been developed for the study of gene function at the molecular level. Initially, the technique involved injecting RNA molecules that were complementary to and would hybridize with particular mRNA molecules for the purpose of disrupting the function of specific genes. Significantly better results have been obtained by injecting double-stranded short *microRNAs* (*miRNAs*, discussed in Chapter 4).

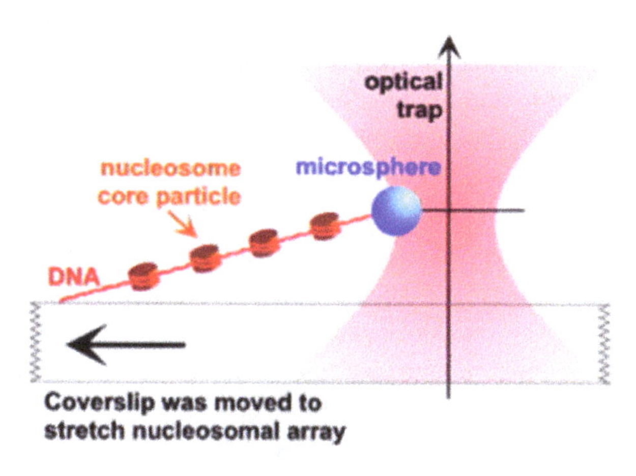

Fig. 8. Single-molecule mechanical manipulations of a chromatin fiber. The fiber consists of a DNA molecule periodically wrapped around nucleosomes; it is attached to a microsphere held by optical tweezers and to a coverslip; it can be stretched by moving the coverslip, and the force necessary to alter its conformation (for example, to unwind it) can be calculated. (Adapted from Brower-Toland *et al.* [4]).

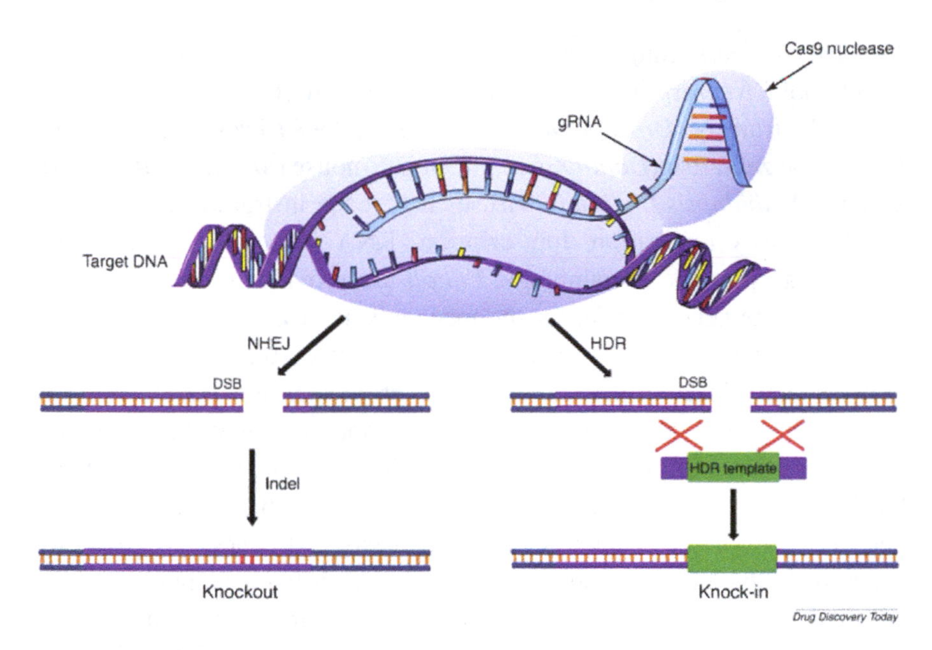

Fig. 9. Diagram illustrating the CRISPR system. A guide RNA molecule (gRNA) contains a sequence to which an enzyme that cuts DNA (Cas9 nuclease) attaches and a sequence that is complementary to a specific region of a gene to be modified. Following the cut, DNA can be repaired by a mechanism that induces a deletion in the gene causing it to lose function (left) or by a mechanism that substitutes information that occurs in the gene with different information. Non-homologous end joining (NHEJ) is a DNA repair mechanism and homology-directed repair (HDR) is usually achieved by homologous recombination when the fragment of homologous DNA is provided. (Adapted from Liu *et al.* [5]).

Another newly invented and amazingly useful biological experimental procedure is the *clustered regularly interspaced short palindromic repeats* (*CRISPR*) gene-editing method. It consists of targeting an enzyme that cuts DNA at a precise site in the genome, most often in order to affect the function of a particular gene (Fig. 9).

D. Model Systems

Through the years, numerous model organisms have been used for genetic research: viruses (T4, ØX, etc.), bacteria (the most commonly used is

Escherichia coli), fungi (the yeast *Saccharomyces cerevisiae* and the bread mold *Neurospora crassa*), the roundworm (*Caenorhabditis elegans*), the fruit fly (*Drosophila melanogaster*), the African frog (*Xenopus laevis*), the zebrafish (*Danio rerio*), the house mouse (*Mus musculus*), and a plant (*Arabidopsis thaliana*). Viruses have been instrumental in understanding DNA replication, transcription, RNA processing, translation, protein transport and immunology. *E. coli* has provided evidence for the validity of the genetic code, for the mechanisms of DNA replication and has allowed the discovery of genetic operons (the co-transcription of several tandemly located genes by a single promoter) and the creation of genetically modified organisms. Neurospora and yeast have been used to connect genes and proteins with their functions in cells; yeast has yielded the basic information on cell cycle regulation and checkpoints that has been key to our understanding of oncogenesis and cancer progression. Caenorhabditis has very rapid generation time, a fixed number of cells whose developmental hierarchy is completely determined, and can be easily grown in mass quantities and provides an excellent model for human aging and neurological diseases. Drosophila is the other invertebrate of choice that has a relatively short life span, has two distinct sexes and has a very small number of chromosomes, and many of its behaviors have been observed in humans. Among vertebrates, zebrafish have a short generation time, produce large numbers of offspring and their embryos are transparent enabling the study of internal structures as they develop. Mice and rats are mammals that share many similarities in anatomy, physiology and genetics with humans. Ultimately, the study of these and many other life forms is performed in order to better understand the developmental biology of humans.

References

[1] Speicher, M.R., and N.P. Carter, The new cytogenetics: blurring the boundaries with molecular biology. *Nature Reviews Genetics*, 2005. **6**: 782–792.

[2] Ambardar, S., *et al.*, High throughput sequencing: an overview of sequencing chemistry. *Indian Journal of Microbiology*, 2016. **56**(4): 394–404.

[3] What is mass spectrometry. The Broad Institute, available at: www.broadinstitute.org/proteomics/what-mass-spectrometry.

[4] Brower-Toland, B., *et al.*, Specific contributions of histone tails and their acetylation to the mechanical stability of nucleosomes. *Journal of Molecular Biology*, 2005. **346**(1): 135–146.
[5] Liu, B., A. Saber, and H.J. Haisma, CRISPR/Cas9: a powerful tool for the identification of new targets for cancer treatment. *Drug Discovery Today*, 2019. **24**(4): 955–970.

CHAPTER 3

The Basic Mechanism of Transcription

Transcription is the mechanism used by a cell to retrieve the information that is present in the genome's DNA. It consists of transferring the code, i.e., the precise sequence of nucleotides of specific regions (transcription units or genes) along the DNA, by synthesizing RNA molecules that are complementary to one strand of these regions: every guanine (G) on the DNA will be represented by a cytosine on the RNA, and vice versa; every thymine (T) on the DNA will be represented by an adenine (A) on the RNA, while an A on the DNA is represented by a new, RNA-specific base — uracil (U) — on the RNA. The sugar in the RNA molecules' sugar–phosphate backbone is ribose. The DNA of eukaryotic organisms includes two different types of genes or transcription units: genes that contain the genetic information for the synthesis of proteins and transcription units that direct the synthesis of different classes of RNA molecules, some of which play a structural role in the assembly of multiprotein components while others participate in or regulate various steps in the information-retrieval process itself. While the mechanisms involved in the transcription of protein-coding and regulatory RNA-coding units exhibit some differences, they are sufficiently similar that the former can be used as a general illustration.

Transcription consists of three separate steps: *initiation*, *elongation* and *termination*. Many transcription units including protein-coding genes are preceded by a region called the *promoter* that straddles the DNA base pair that is the first to be transcribed, designated as the *transcription start site* or TSS [1]. To initiate transcription, a large molecular complex, the *pre-initiation complex* (PIC), is formed containing the enzyme necessary to synthesize an RNA that is the exact copy of one of the strands of the DNA transcription unit. The promoter region contains several DNA elements that facilitate the binding of various components of the PIC. In the case of protein-coding genes and of some regulatory RNA transcription units, the enzyme responsible for RNA synthesis is *RNA polymerase II* (RNAPII). Other regulatory RNAs are transcribed by different polymerases (RNAPI or III). The RNA produced by the transcription of protein-coding genes is referred to as *messenger RNA* (mRNA). All of these polymerases are large enzymes made up of multiple subunits. In addition to the RNA polymerase enzyme, the PIC includes components — *transcription factors* (TFs)[1] — that interact with other factors present on or near the promoter or that recognize and bind to the promoter region. The initiation of the transcription process, i.e., RNA synthesis, almost always requires the association of the PIC bound to the promoter with activation factors present at additional DNA regulatory modules (*enhancers*) located on either side of the transcription units, sometimes within, but often at substantial distances from the units (Fig. 1). The association of enhancers and promoters is stabilized by a molecular complex described in Chapter 5. The activation factors contributed to the PIC by the enhancer are regulatory proteins that specify gene activation in relation to the cell cycle, as a response to stress signals or other environmental clues, or to confer tissue specificity [3]. Enhancers usually interact with the PIC through a large multiprotein complex termed the *Mediator complex* that transmits signals from transcription factors to RNAPII through mechanisms that are still poorly understood. The physical association of enhancer and promoter complexes is achieved by physical looping and is maintained by special proteins that form the *cohesin* complex [4].

---

[1]The six basic transcription factors (TFIID, IIA, IIB, IIE, IIF and IIH) perform different functions. TFIID binds to the promoter; this association requires TFIIA and is positioned correctly with respect to the TSS by TFIIB and TFIIF; TFIIE serves as a bridge between RNAPII and these factors; TFIIH contains a helicase — an enzyme that unwinds DNA — that is regulated by TFIIE. In addition to these general factors, a number of additional factors (TAFs) are involved in the formation of the PIC.

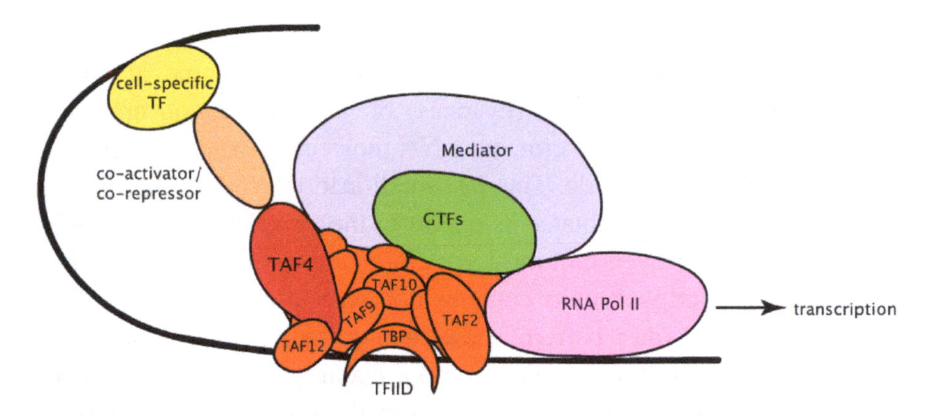

Fig. 1. Diagram of gene transcription. The pre-initiation complex includes the RNA polymerase II enzyme and interacts with transcription factors (GTFs and TAFs) via the Mediator complex (from Kazantseva and Palm [2]).

To initiate transcription, (i) the two strands of the DNA unit to be transcribed must be separated, which will allow the synthesis of an RNA molecule that is complementary to the base sequence of one strand called the *template* or *sense strand* and, thereby, will reflect the nucleotide sequence of the DNA in that region (separation of the DNA strands is achieved by the action of *helicases*, enzymes that are present in the PIC), and (ii) a major subunit of RNAPII must be modified (phosphorylated). The RNAPII and some of the transcription factors present in the PIC, now referred to as the *transcription elongation complex* (TEC), leave the promoter and begin the synthesis of the RNA. Other factors and the Mediator complex are left on the promoter to participate in the formation of the next PIC. As elongation proceeds, the area of separation of the two complementary DNA strands — the *transcription bubble* — moves along with the TEC. In the case of most protein-coding genes in eukaryotes, the process of transcription stops after the synthesis of a 20- to 30-nucleotide-long RNA [5]. This pausing of the TEC, induced by specific factors, and its subsequent resumption are an important means of transcription regulation [6].

Several events mark the resumption of mRNA synthesis by the TEC [7]. The RNAPII major subunit must be further modified (once again by phosphorylation) and the factors that caused it to pause must be inactivated.[2] Elongation involves transcription factors, some of which modify the chromatin proteins (histones) of the transcription unit, and capping of the 5′ end of the growing RNA molecule by the addition of a modified guanine nucleotide. This last modification protects the mRNA from degradation and facilitates its export to the cytoplasm where it will be translated. The sequence of most protein-coding genes and some regulatory RNA transcription units contains nonsensical stretches of nucleotides (*introns*) that must be removed allowing the meaningful stretches (*exons*) to be concatenated (*spliced*) into a mature, functional transcript. Very frequently, some exons are skipped in the process of producing a functional transcript. This process, referred to as *alternative splicing*, results in the production of different proteins from a single gene (Fig. 2). Obviously, the transcription of a given unit must be terminated to prevent the TEC from including the sequence of the adjacent unit in the RNA that it synthesizes. This is achieved by the presence of special nucleotides, termed transcription *termination* sequences, that recruit factors responsible for releasing the polymerase from the DNA template. Lastly, in most organisms, a series of adenine nucleotides are added to the 3′ end of the transcript (*polyadenylation*) that will protect it from intracellular degradation. Once synthesized and modified, the mature mRNA exits the cell nucleus and provides the template for the synthesis of a protein.

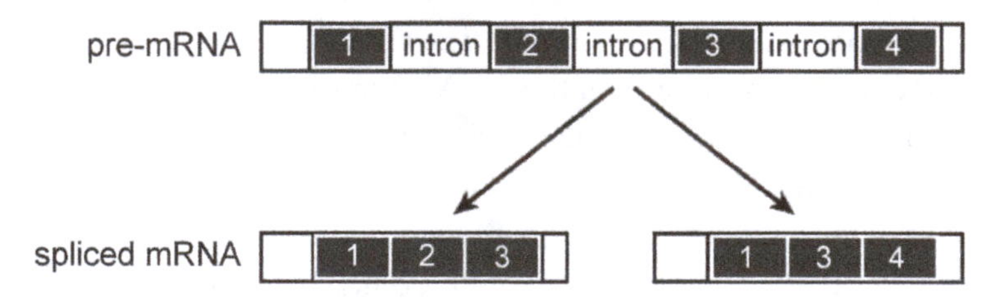

Fig. 2. Diagram illustrating the alternative splicing of a primary transcript to yield two different mRNA molecules that would result in two different proteins.

[2]The two factors involved in RNAPII pausing are *DRB sensitivity inducing factor* (DSIF) and *negative elongation factor* (NELF). DRB stands for 5,6-dichloro-1-β-D-ribofuranosylbenzimidazole, a chemical compound that inhibits transcription elongation. Release from pausing involves the phosphorylation of DSIF and NELF; NELF dissociates from the early elongation complex and phosphorylated DSIF attracts factors that stimulate productive transcript synthesis.

Box 1

Translation of mRNA and protein synthesis: Translation is the process used by cells to transform the genetic code contained in mRNA molecules into proteins. This process is carried out by *ribosomes*, complex molecular machines made up of several RNA molecules (rRNAs) and numerous proteins subdivided into two subunits; in eukaryotes, the small subunit contains one RNA molecule and 33 different proteins, while the large subunit contains three additional RNA molecules and 46 proteins. Ribosomes are assembled in the nucleolus (see Chapter 5). The ribosomes' function is to ensure that transfer RNA (tRNA) molecules containing the appropriate amino acids bind to the codons in the mRNA that specify these amino acids. Since there are 64 possible combinations of the four nucleotides taken three at a time, and only 20 different amino acids, some amino acids are specified by more than one codon (see Chapter 1, Box 2). Each of the different tRNA molecules present in a cell can attach to a particular amino acid and has a region (anti-codon) that can base-pair with a particular codon in the mRNA; this codon calls for that amino acid to be present in the protein that is being synthesized (Fig. 3).

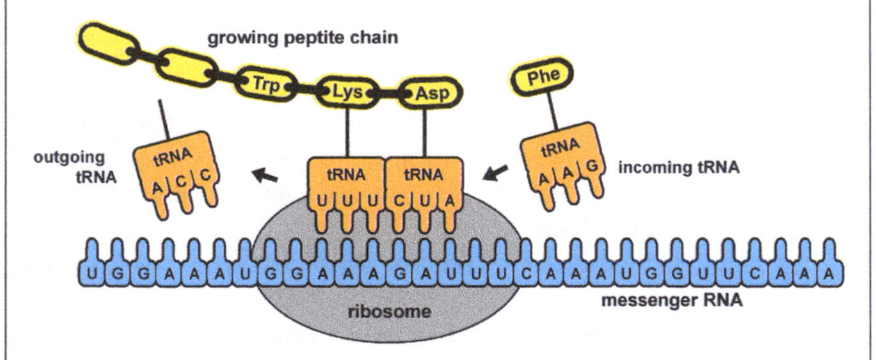

Fig. 3. Translation of mRNA molecules. In the first step, a tRNA molecule carrying a particular amino acid and the appropriate anti-codon binds to the mRNA codon. As this amino acid becomes linked (by a peptide bond) to the amino acid brought by the previous tRNA, this tRNA is released. The ribosome slides and the next codon is now available for a new tRNA to bind to it.

This process, termed *translation*, is carried out in the cytoplasm by multiprotein particles: the *ribosomes* (see Box 1). It results in the formation of a chain of amino acids that corresponds to the chain of coding nucleotide triplets in the mRNA (Fig. 4).

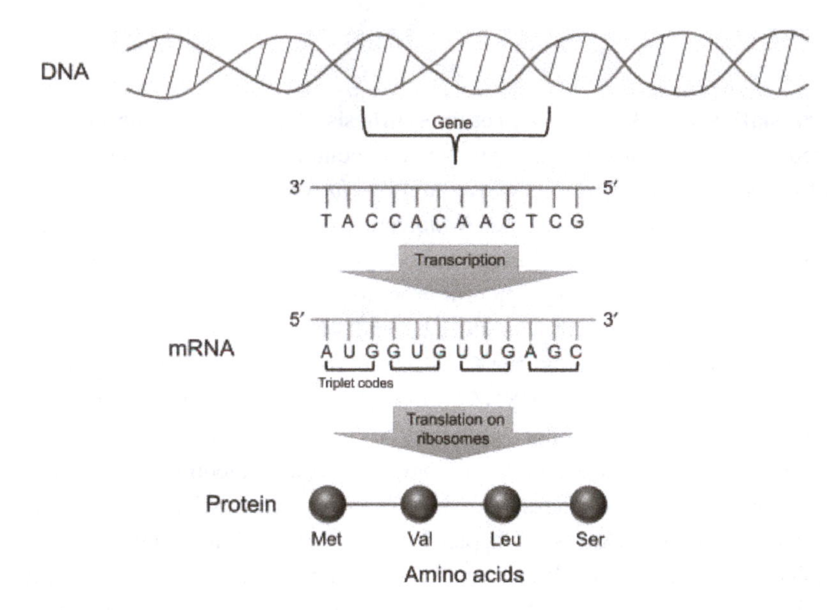

Fig. 4. Diagram illustrating the flow of genetic information from a protein-coding gene to the protein that it encodes (from Chang-Hui Shen [8]).

References

[1] Riethoven, J.-J.M., Regulatory regions in DNA: Promoters, enhancers, silencers, and insulators. *Methods in Molecular Biology*, 2010. **674**: 33–42.

[2] Kazantseva, J., and K. Palm, Diversity in TAF proteomics: Consequences for cellular differentiation and migration. *International Journal of Molecular Science*, 2014. **15**(9): 16680–16697.

[3] Bulger, M., and M. Groudine, Functional and mechanistic diversity of distal transcription enhancers. *Cell*, 2011. **144**(3): 327–339.

[4] Kagey, M.H., *et al.*, Mediator and cohesion connect gene expression and chromatin architecture. *Nature*, 2010. **467**(7314): 430–435.

[5] Rougvie, A.E., and J.T. Lis, The RNA polymerase II molecule at the 5′ end of the uninduced hsp70 gene of *D. melanogaster* is transcriptionally engaged. *Cell*, 1988. **54**(6): 795–804.

[6] Nechaev, S., and K. Adelman, Pol II waiting in the starting gates: Regulating the transition from transcription initiation into productive elongation. *Biochimica et Biophysica Acta*, 2011. **1809**(1): 34–45.

[7] Bentley, D.L., Coupling mRNA processing with transcription in time and space. *Nature Reviews Genetics*, 2014. **15**(3): 163–175.

[8] Shen, C.-H., *Diagnostic Molecular Biology*, 2019. Academic Press.

CHAPTER 4

Chromatin Structure

A. Nucleosomes and Assembly of the Chromatin Fiber

Histones and the organization of the nucleosome: In the nuclei of cells, the genetic material is present in the form of chromatin, a complex made up of the DNA and different types of RNAs and proteins that include the histones. The basic unit of chromatin in all eukaryotic organisms is called the *nucleosome*; it consists of a segment of DNA wrapped around an octamer made up of two copies of the four core histones: H2A, H2B, H3 and H4 (Fig. 1). The amino acid sequences of the core histones are among the most conserved in all eukaryotes; because during cell division they form the nucleosomes that become associated with the newly replicated DNA, they are referred to as *replication-coupled* (RC) histones. When cells perform specific functions such as transcription or the repair of DNA damage, or in the case of some specific cell types such as sperm cells, some of the core histones are replaced by variants that are different in their amino acid sequence [1]. These variants, said to be *replication-independent* (RI), are introduced when nucleosomes are evicted during transcription (Chapter 6) or as a result of DNA damage (Chapter 8).

A fifth histone, the *linker histone* (called H1, H5, etc. depending on the particular organism) associates with the DNA that is wrapped around

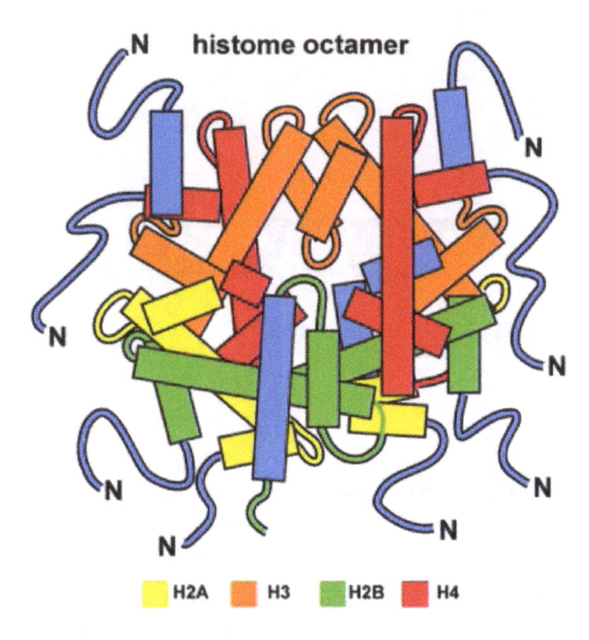

Fig. 1. Diagram illustrating the structure of a histone octamer. The cylindrical parts indicate the regions that cause the different histone molecules to fold and assume their characteristic shape. The N-terminal strands are the histone tails where most of the chemical modifications that affect gene activity are located.

the core histone octamer and confers stability to the nucleosome. This histone is normally removed in order to allow gene activation (see Box 1).

DNA sequence and nucleosome positioning: In general, the DNA molecule makes 1.65 turns around the histone octamer with which it has a number of contacts based on differences in charge of their respective ions. The length of DNA in a nucleosome is 147 nucleotide pairs, while the segment of DNA between adjacent nucleosomes — *linker* DNA — can vary in length and ranges, on average, between 20 and 60 nucleotide pairs in different tissues or species (Fig. 3).

During cell division, the replication of the parent cell's DNA into the two copies that are destined for the *daughter* cells requires the removal of nucleosomes. Following replication, the histones that were present on the original DNA molecule are randomly distributed to the nascent DNA and

Box 1

Linker histones: These histones are responsible for compacting the chromatin fiber. Linker histones have a tripartite structure consisting of an N-terminal and a C-terminal domain separated by a globular region. A linker histone molecule binds to the DNA that is looped around the nucleosome and to the DNA as it enters and exits the nucleosome, bringing these two DNA regions closer together (Fig. 2). Although their structure is highly conserved, the amino acid sequence of linker histones varies significantly from organism to organism; in addition, some organisms such as humans have multiple variants, some of which are found in somatic cells while others are specific of male or female germ cells.

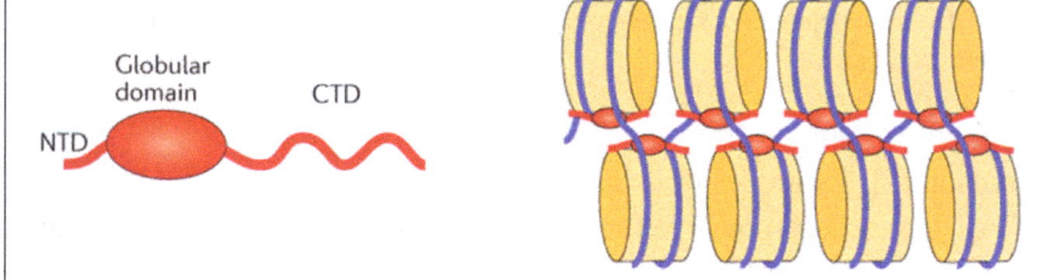

Fig. 2. Structure (left) and attachment (right) of linker histones to nucleosomes (modified from Fyodorov *et al.*, 2019).

The role that linker histones play in regulating gene function depends on their covalent modifications which, as in the case of the core histones, can either activate or repress transcription. The presence of linker histones also has a negative effect on DNA replication and DNA repair. To allow both of these activities to occur, chromatin undergoes substantial reorganization that includes the eviction of linker histones.[a]

[a]Fyodorov, D.M., *et al.*, Emerging roles of linker histones in regulating chromatin structure and function. *Nature Reviews Molecular Cell Biology*, 2019. **19**(3): 192–206.

new core histones are synthesized to restore the full nucleosome complement on each new DNA molecule. New and old histones associate with specific protein factors called *chaperones* that deposit them on the newly synthesized DNA copies to reconstitute the nucleosomal architecture of

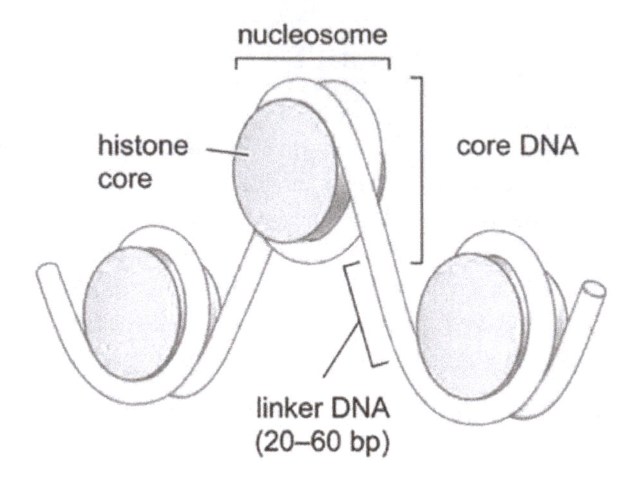

Fig. 3. Diagram depicting a segment of a DNA molecule wrapped around three successive nucleosomes.

chromatin [2]. Examples of chaperones are the *chromatin assembly factor 1* (CAF-1) that is responsible for the deposition of histones H3 and H4 and *nucleosome assembly protein 1* (Nap1) that delivers histones H2A and H2B. Core histones are replaced by variant histones through the action of other dedicated chaperones. The positioning of nucleosomes along the DNA, although influenced by the DNA sequence (some regions appear to have greater and others lesser affinity for nucleosomes), is mostly determined by the action of *remodeling factors*. Examples of some of these factors are assembly of *core histones factor* (ACF), *chromatin accessibility complex* (CHRAC), *nucleosome remodeling factor* (NURF) and *remodeling and spacing factor* (RSF). These factors usually direct the formation of regularly spaced nucleosomes that may result in greater compaction of the chromatin fiber [3]. The progress of RNA polymerases during the process of transcription and of DNA polymerases during the process of DNA replication is impeded by the presence of nucleosomes. Therefore, other remodeling factors exist that move or eject nucleosomes (see the following). In the nuclei of cells, chromatin assumes the configuration of a series of beads on a string of 10 nanometers (nm) in width. By changing the buffer to more physiological conditions, the fiber condenses into a secondary structure that is 30 nm wide (Fig. 4). It was thought for some time that

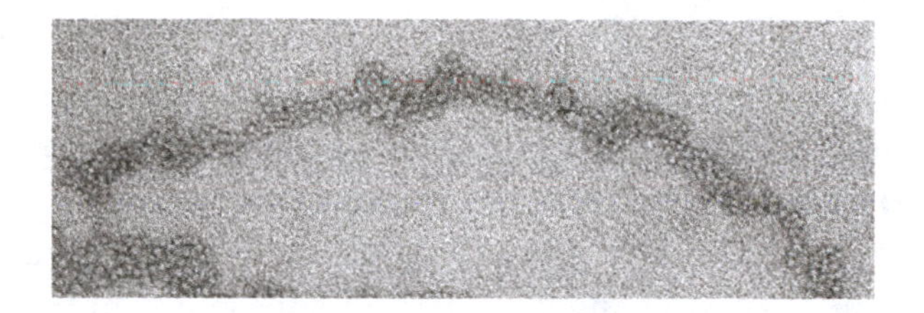

Fig. 4. Electron photomicrographs illustrating the 30 nm chromatin fiber. The width of the fiber corresponds to approximately 2 to 3 nucleosomes (modified from Rattner and Hamkalo [4]).

this was the natural state of chromatin in living cells. More recent observations have led to the suggestion that chromatin exists in nuclei as irregularly folded, sometimes globular structures of 10 nm fibers [5].

B. The Different States of the Chromatin Fiber

Euchromatin and heterochromatin: There are four major reasons for the highly folded architecture of chromatin: (i) to accommodate very long DNA molecules into the nucleus between cell divisions, (ii) to facilitate the distribution of these molecules following DNA replication to daughter cells, (iii) to protect the DNA from harmful chemical or physical damage and (iv) to allow the orderly retrieval of selected genetic information. Regarding the latter, chromatin is different in genetically active and inactive regions and is referred to as *euchromatin* and *heterochromatin*, respectively. Some heterochromatic regions of chromosomes remain condensed at all times and are referred to as *constitutive heterochromatin*; other regions are inactive at some time or in some cell types but not in others and are referred to as *facultative heterochromatin*.

Nuclear distribution of the different chromatin states: Other distinctions between active and inactive chromatin are their location within the nucleus and the timing of their replication in preparation for cell division. Regions of condensed chromatin tend to be associated with the nuclear periphery while more diffuse chromatin has a more central localization. This difference plays a role in regulating the transcription process. Heterochromatic regions lag behind during genome duplication.

The molecular formation and constitution of heterochromatin were elucidated by studying the genes responsible for a genetic phenomenon called *position effect variegation* [6]. This phenomenon manifests itself when a wild-type gene is relocated by some chromosome rearrangement to the vicinity of a block of constitutive heterochromatin; the gene is inactivated in some tissues, giving a variegated appearance to the organism (*variegated phenotype*). These studies uncovered a number of heterochromatin-specific proteins and chromatin-modifying enzymes.

C. Modulating the Structure of Chromatin to Regulate Transcription

As mentioned above, a major role for the organization of the genome into chromatin is that, although it exists in a fairly unraveled condition during the major part of the cell cycle, it can be highly condensed to allow the proper distribution of the genetic material to daughter cells following cell division. The other major role is to regulate the retrieval and function of the genetic information encoded in the DNA.

Post-translational covalent modifications of histones: RNA polymerase is unable to transcribe DNA that is wrapped around nucleosomes because promoters and enhancers are unavailable to the various transcription factors that need to bind to these regulatory regions, and the compaction of inactive chromatin is refractory to the association of large regulatory complexes, such as the pre-initiation complex. Therefore, the activation of genes, i.e., the onset of transcription, requires alterations in the structure of chromatin [7]. An amazing variety of activities have evolved to achieve this purpose, ranging from the covalent modification of histone molecules or of DNA bases and the displacement of nucleosomes by multiprotein complexes, to the use of small or long non-coding, regulatory RNAs. These modifications do not affect the coding sequence of the DNA; therefore, rather than being considered "genetic" in nature, they are said to be "epigenetic". Not surprisingly, a reversal of these alterations is necessary for silencing genes or maintaining regions of chromatin in a repressed state.

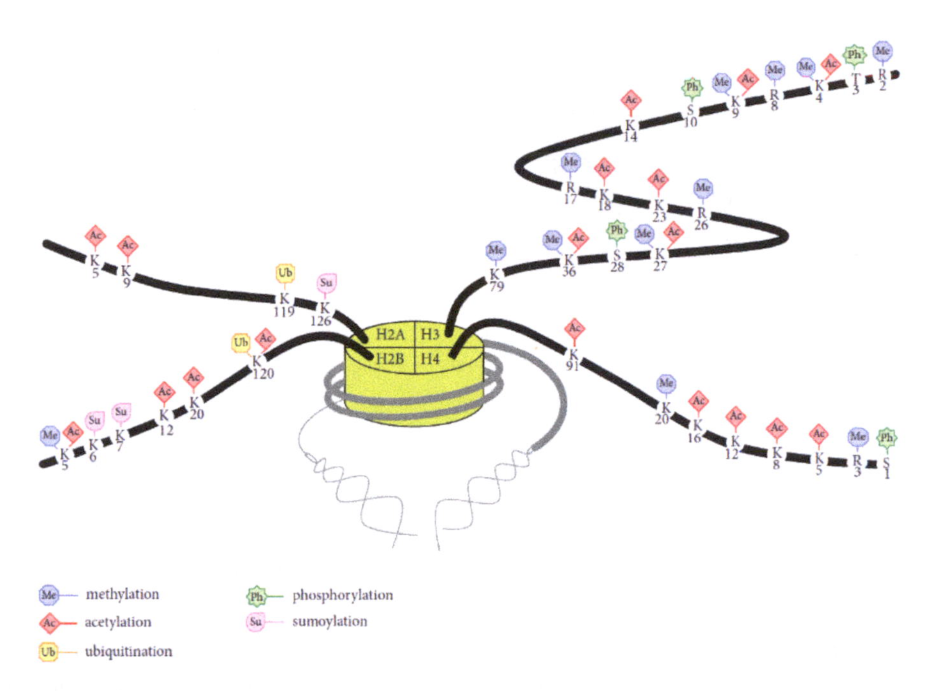

Fig. 5. Possible post-translational modifications of the core histones. The letter code for the different amino acids is as follows: K = lysine, R = arginine, T = tyrosine and S = serine (from Araki and Mimura [8]).

Covalent modifications of histones: Most of these modifications occur in the tails that extend from the globular core of the histone octamers; others occur at specific sites throughout the length of particular histones (Fig. 5). Because they are added to the histones' molecular structure after their synthesis, they are referred to as *post-translational modifications* (PTMs). Some of the most studied histone modifications are *acetylation, methylation, phosphorylation* and *ubiquitination*; although many other modifications have been identified, these four have major influence on transcription and silencing [9]. The amino acid lysine (letter code K) can be acetylated by histone lysine acetyltransferase enzymes (HKATs) that add a CH_3CO group, using acetyl-coenzyme A (acetyl-CoA) as the donor; this reaction can be reversed by the action of deacetylases (HKDACs) (Fig. 6). Lysine acetylation neutralizes the positive charge on this amino

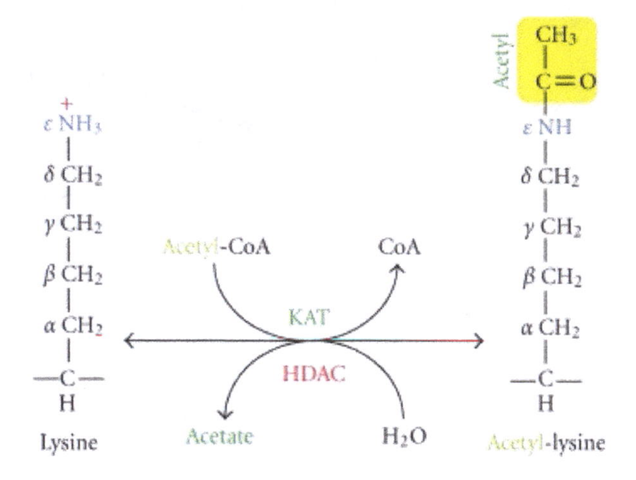

Fig. 6. Acetylation and deacetylation of lysine (from Kim *et al.* [10]).

acid and, thereby, reduces the attraction of the negatively charged DNA molecule to the modified histone.

Because of this general effect, particular acetylated lysines perform very specific roles in regulating gene function: they loosen the condensation of chromatin or directly prevent it from becoming silent. In contrast, the acetylation of some other lysines is associated with the binding of complexes that repress transcription. In addition to acetylation, a number of different *acylations* of histone lysines can occur (crotonylation, formylation, butyrylation, succinylation and malonylation). These acylation reactions use different forms of coenzyme-A as a donor and, similarly to acetylation, alter the charge of the lysine residue.

Lysine residues can be methylated by histone lysine methyl transferases (HKMT) that can add up to three methyl groups, using S-adenosylmethionine (SAM) as a donor (Fig. 7). Unlike acetylation, this modification does not alter the charge of the histone molecule but serves to attract complexes that regulate chromatin function. Depending on their position along the histone tail, methylated lysines can be associated with gene activation or repression. The methyl groups can be removed by lysine-specific demethylases (LSD). Arginine (R) is another amino acid in histones that can be methylated. Usually, this modification affects the ability of neighboring amino acids to be modified.

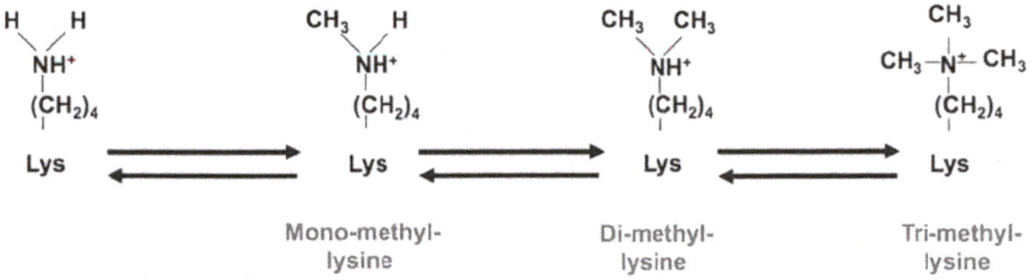

Fig. 7. Lysine methylation and demethylation.

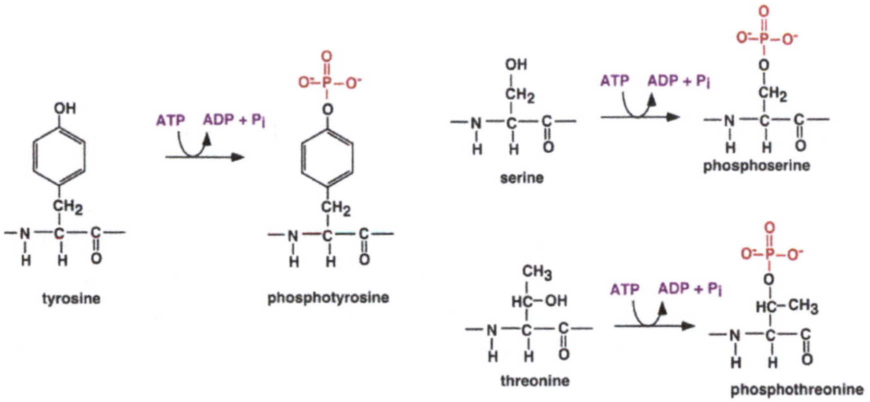

Fig. 8. Phosphorylation of histone residues.

Three histone amino acids, serine (S), threonine (T) and tyrosine (Y), can be phosphorylated by different kinases that use adenosine triphosphate (ATP) as the phosphate group donor (Fig. 8). As for acetylation, phosphorylation of histones reduces their association with DNA, facilitating gene transcription. Surprisingly, when the cell is preparing to enter mitosis, extensive histone phosphorylation is necessary for the process of chromatin condensation.

Ubiquitination consists of the addition of ubiquitin, a 76-amino-acid-long peptide, to a lysine residue. Three enzymatic reactions are involved: the first one uses the hydrolysis of adenosine triphosphate (ATP) to generate the energy necessary for activation of the ubiquitin peptide and to deliver it to a conjugating enzyme; the ubiquitin is then transferred to a

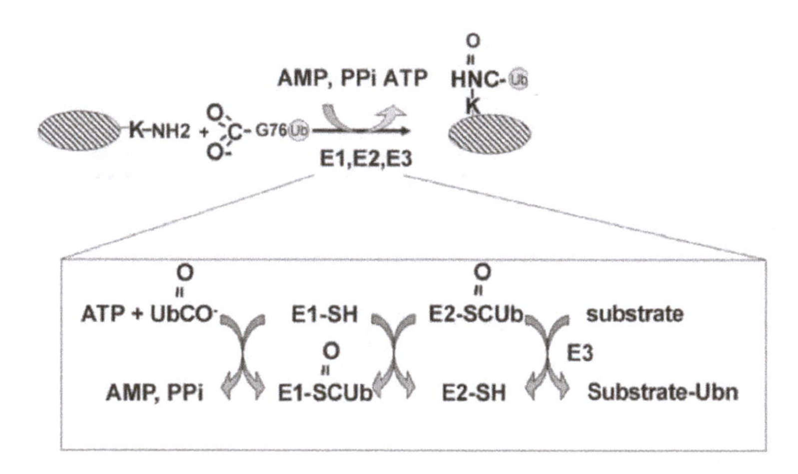

Fig. 9. Reactions involved in lysine ubiquitination (from Jadhav and Wooten [11]).

third enzyme that links it to the histone molecule (Fig. 9). The addition of ubiquitin to lysines on different histone molecules has opposite effects on transcription: on histone H2B, its presence is necessary for the methylation of other lysines that are associated with active transcription; in contrast, the ubiquitination of H2A is required for gene silencing.

The addition of *sumo* — a peptide that resembles ubiquitin — to lysines, on all four core histones, is achieved by the successive activity of three enzymes, in a manner similar to those responsible for ubiquitination. Sumoylation is associated with the repression of transcription.

A number of additional covalent modifications of histones have been reported. Among them are *glycosylation*, the addition of β-N-acetyl-glucosamine residues (O-GlcNAc) that appear to be involved in nucleosome assembly as well as the modulation of transcription, *ADP ribosyla*tion, the addition of single or multiple ADP-ribose moieties, thought to diminish the contacts between histones and DNA, and *hydroxyisobutyrylation* of lysine that can occur on all core histones, once again implicated in active transcription.

Covalent modifications of DNA: DNA modifications consist predominantly of the methylation or hydroxymethylation of cytosines, usually when they are followed by a guanine (represented as CpG dinucleotides, where p is the sugar–phosphate link between neighboring bases). Methylation is the responsibility of *DNA methyltransferases* (DNMTs) that use S-adenosylmethionine as the donor of the methyl group. Two of these enzymes, *de novo* DNMTs, add new methyl groups to DNA and the third, *maintenance* DNMT, maintains the existing methylation on newly replicated DNA. The removal of methylation can occur either enzymatically or simply by the failure to maintain it during DNA replication. Hydroxymethylcytosine is derived from methylcytosine.

DNA methylation is associated with regions or entities of the genome that must remain silenced. Such regions consist of *repetitive sequences* present in tandem arrays, i.e., that follow one another in succession in the chromosomal DNA, and *transposable elements*. The former are prevalent around the centromere region of chromosomes; the latter are mobile genetic elements that constitute a major fraction of eukaryotic genomes; these elements can relocate (*transpose*) within the genome and thereby cause mutations by interfering with the function of genes at or near their new location; such transpositions are prevented if the mobile elements' DNA is methylated. *Imprinted* genes represent another case of permanent silencing that involves DNA methylation. Only one of the two alleles of a few genes in the genome of mammals, either the allele inherited from the mother or the one inherited from the father, is expressed in the developing embryo and the resulting adult; in most cases, the allele that remains silenced is marked by DNA methylation. This mark constitutes an imprint that attracts other factors and modifications characteristic of inactive chromatin [12]. The explanation for the existence of imprinted genes remains elusive.

DNA methylation is also strongly implicated in the activation or repression of genes during cellular differentiation and function. The dynamics of this type of gene regulation are variable and therefore, are quite complex [13]. For example, while regulatory regions (enhancers and promoters) are usually demethylated as they are bound by transcription factors, the methylation of gene bodies increases with transcription. As expected, there is a correlation between DNA methylation and histone modifications: modifications that are involved in gene repression are found in methylated DNA regions of the genome; modifications that are associated with gene activation appear to prevent DNA methylation. This renders the methylation that occurs along gene bodies somewhat puzzling. Possible explanations include the possibility that it facilitates the process of transcription elongation and that it prevents the ectopic initiation of transcription at internal sites within genes.

Remodeling complexes: These large multiprotein complexes alter the position of nucleosomes along the DNA, evict the histone octamers or modify the association of the DNA with the octamers (Fig. 10). These actions can uncover promoter regions that are otherwise unavailable to transcription factors and activators. Remodelers belong to four families: *switching defective/sucrose non-fermenting* (SWI/SNF), imitation switch (ISWI), *inositol requiring 80* (INO80) and *chromodomain helicase DNA-binding* (CHD) complexes. Some complexes (SWI/SNF and CHD) slide and eject nucleosomes, some alter the distance between nucleosomes (ISWI) and some are involved in replacing core histones with variants (INO80). Other remodeling complexes are involved in gene repression; an example is the *nucleosome remodeling and deacetylase* (NuRD) complex.

Complexes that are highly related to these and others exist in all eukaryotes. To perform their functions, all complexes use the energy liberated by the hydrolysis of ATP [15].

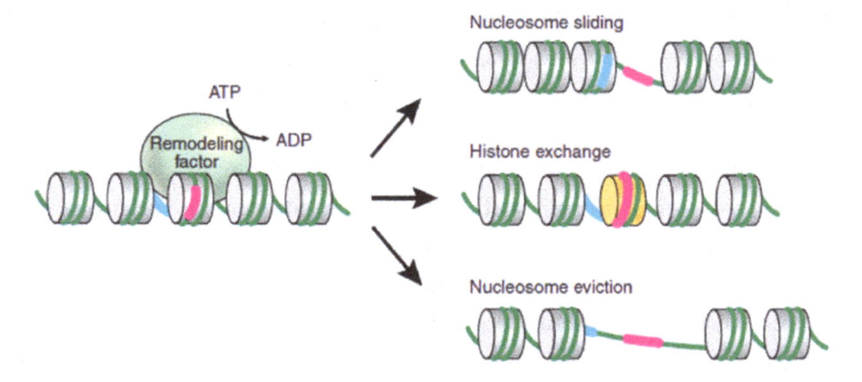

Fig. 10. ATP-dependent chromatin remodeling complexes slide or evict nucleosomes to expose previously masked DNA sequences; they can also replace core histones with variants (modified from Becker and Workman [14]).

Histone variants: As mentioned above, some complexes replace core histones with variants. For example, the histone H3.3 variant differs from H3 in four amino acids. This variant is deposited into newly formed octamers, in active gene sites by some remodelers and at inactive gene sites by others, where it functions to prevent or contribute to gene silencing. Histone H2A has many variants that participate in the repair of DNA damage (H2A.X) or stimulate transcription (H2A.Z). In summary, histone variants greatly expand the function of nucleosomes by changing the level and stability of DNA wrapping or by interacting with other chromatin components.

D. Non-coding RNAs and RNA Editing

In many organisms, a very small percentage of the genome of eukaryotes consists of protein-coding genes (75% in yeast and only 2% in humans). Most of the remainder of genomes consist of transcription units that encode RNA molecules of different types; these RNAs play indispensable roles in the retrieval and use of the genetic information. Two major classes of non-coding RNAs are involved in the process of translation, i.e., the synthesis of proteins, that takes place in the cellular cytoplasm: *ribosomal RNAs* (rRNAs) that are the structural components of ribosomes and *transfer RNAs* (tRNAs) that carry amino acids to the specific codons in the messenger RNA as it is translated by the ribosomes (see Chapter 2, Box 1). The third major class consists of many different types of RNAs belonging to two general groups — small and long non-coding RNAs — most of which remain within the nucleus and regulate gene activity.

Small non-coding RNAs (snRNAs): Several different types of these RNAs exist with different functions [16–18]. *MicroRNAs* (miRNAs) are usually around 22 nucleotides long; they are produced in the nucleus and are assembled into ribonucleoprotein complexes in the cytoplasm where they regulate the level of particular gene transcripts by associating with the mRNA molecules either directing their degradation or preventing their use in protein synthesis. *Small nuclear RNAs* (snRNAs) are around 150 nucleotides in length; most of them are components of RNA–protein complexes (*small nuclear ribonucleoproteins* or snRNPs) that combine with other proteins to form *spliceosomes*: large complexes responsible for splicing pre-mRNA molecules. *Small nucleolar RNAs* (snoRNAs), 60 to 170 nucleotides in length, participate in the function of complexes that modify some of the nitrogenous bases of ribosomal RNAs (RNA editing; see the following). *PIWI-interacting RNAs* (piRNAs) associate with PIWI proteins (PIWI is an acronym for the first example discovered in *Drosophila*). These piRNAs occur only in animals where one of their main functions is to defend against the movement of transposable elements. Lastly, *tRNA-derived small RNAs* (tsRNAs), produced by the fragmentation of tRNAs, play an important role in regulating translation by targeting specific mRNAs. Several other types of small non-coding RNAs have been identified and their respective activities are currently being investigated.

Long non-coding RNAs (lncRNAs): These RNAs are over 200 nucleotides and can reach much greater lengths. Most of them undergo splicing and exert their functions within the nucleus; some are processed in a different manner and function in the cytoplasm [19]. Some of these RNAs modulate the function of nearby genes or entire chromosomes by spreading along the chromosome from the site of their transcription and recruiting regulatory factors (see Chapter 6). Other lncRNAs may affect the expression of adjacent genes because they partially overlap and their transcription interferes with the transcription of the genes. In some cases, regulatory sequences within the lncRNA gene and not the transcription product of the gene, i.e., the lncRNA, appear to have a regulatory effect on nearby genes, acting in a similar manner as enhancer elements. An increasing number of lncRNAs that affect chromatin structure and gene expression in regions of the genome that are distant from their site of transcription are being identified. Some of these RNAs act by binding to distant genes and recruiting regulatory proteins; some bind to particular miRNAs preventing the latter's inhibitory action; others appear to be responsible for the clustering of distant genes for regulatory purposes. Many lncRNAs participate in the formation of some nuclear bodies — regions of molecular condensates that are not surrounded by membranes and are present in the nuclei of eukaryotic cells where many biochemical processes are carried out (see Chapter 5). Finally, some lncRNAs leave the nucleus and either enhance or prevent the translation of particular mRNAs.

Enhancer RNAs (eRNAs): This is a large group of non-coding RNAs transcribed from enhancer sequences in a manner that is very similar to the transcription of the protein-coding genes and lncRNAs, although the majority of eRNAs are not polyadenylated or spliced [20]. Some eRNAs act on genes that are adjacent to the enhancer region where they were transcribed; others relocate to distant genomic regions. In both cases, these RNAs affect gene activity by interacting with histone-modifying factors that relax chromatin or by associating with RNAPII and various transcription factors.

RNA editing: All of the different types of RNAs can undergo a variety of covalent modifications. The most common are methylation of all four nitrogenous bases and of the ribose sugar, the conversion of adenine to hypoxanthine (hypoxanthine attached to ribose is known as inosine), the conversion of uridine to its isomer pseudouridine and the conversion of cytosine to uracil [21]. Many mRNAs (and some lncRNAs and miRNAs) have multiple methylated adenosines. Some of these methylations can target the mRNAs for degradation, others can promote their translation into proteins. The RNAs involved in the translation process, tRNAs and rRNAs, are extensively modified; the modifications optimize their role in translation.

References

[1] Talbert, P.B., and S. Henikoff, Histone Variants at a glance. *Journal of Cell Science*, 2021. **134**(6): 1–13.

[2] Hammond, C.M., *et al.*, Histone chaperone networks shaping chromatin function. *Nature Reviews in Molecular Cell Biology*, 2017. **18**(3): 141–158.

[3] Baldi, S., *et al.*, Genome-wide rules of nucleosome phasing in *Drosophila*. *Molecular Cell*, 2018. **72**(4): 661–672.

[4] Rattner, J.B., and B.A. Hamkalo, Nucleosome packing in interphase chromatin. *Journal of Cell Biology*, 1979. **81**(2): 453–457.

[5] Maeshima, K., S. Ide, and M. Babokhov, Dynamic chromatin organization without the 30-nm fiber. *Current Opinion in Cell Biology*, 2019. **58**: 95–104.

[6] Elgin, S.C.R., and G. Reuter, Position effect variegation, heterochromatin formation, and gene silencing in Drosophila. *Cold Spring Harbor Perspectives in Biology*, 2013. **5**(8): 1–26.

[7] Eissenberg, J.C., and J.C. Lucchesi, Chromatin structure and transcriptional activity of an X-linked heat shock gene in *Drosophila pseudoobscura. The Journal of Biological Chemistry*, 1983. **258**(22): 13986–13991.

[8] Araki, and T. Mimura, The histone modification code in the pathogenesis of autoimmune diseases. *Mediators of Inflammation*, 2017. Article ID 2608605, 12 pages.

[9] Talbert, P.B., and S. Henikoff, The yin and yang of histone marks in transcription. *Annual Review of Genomics and Human Genetics*, 2021. **22**: 147–170.

[10] Kim, G.-W., *et al.*, Dietary, metabolic and potentially environmental modulation of the lysine acetylation machinery. *International Journal of Cell Biology*, 2010. Article ID 632739, 14 pages. 11. Jadhav, T., and M.W. Wooten, Defining an embedded code for protein ubiquitination. *Journal of Proteomics and Bioinformatics*, 2009. **2**(7): 316–333.

[12] Ferguson-Smith, A.C., and D. Bourc'his, 2018 Gairdner Awards: The discovery and importance of genomic imprinting. *eLife*, 2018. Article ID 7:e42368, 5 pages.

[13] Ginno, P.A., *et al.*, A genome-scale map of DNA methylation turnover identifies site-specific dependencies of DNMT and TET activity. *Nature Communications*, 2020. **11**(1): 1–16.

[14] Becker, P.B., and J. Workman, Nucleosome remodeling and epigenetics. *Cold Spring Harbor Perspectives in Biology*, 2013. **5**(9): 1–19.

[15] Clapier, C.R., and B.R. Cairns, The biology of chromatin remodeling complexes. *Annual Review of Biochemistry*, 2009. **78**: 273–304.

[16] Zhang, P., *et al.*, Non-coding RNAs and their integrated networks. *Journal of Integrative Bioinformatics*, 2019. **16**(3): 1–12.

[17] Ozata, D.M., *et al.*, PIWI-interacting RNAs: small RNAs with big functions. *Nature Reviews Genetics*, 2019. **20**(2): 89–108.

[18] Shi, J., *et al.*, tsRNAs: The Swiss army knife for translational regulation. *Trends in Biochemical Sciences*, 2019. **44**(3): 185–189.

[19] Satello, L., *et al.*, Gene regulation by long non-coding RNAs and its biological functions. *Nature Reviews Molecular Cell Biology*, **22**(2): 96–118.

[20] Sartorelli, V., and S.M. Lauberth, Enhancer RNAs are an important regulatory layer of the epigenome. *Nature Structural and Molecular Biology*, 2020. **27**(6): 521–528.

[21] Willbanks, A., S. Wood, and J.X. Cheng, RNA epigenetics: Fine-tuning chromatin plasticity and transcriptional regulation, and the implications in human diseases. *Genes*, 2021. **12**(5): 1–41.

CHAPTER 5

Nuclear Organization

A. Architectural Organization of the Genome

Chromosome territories and compartments: In eukaryotic organisms, when cells are not dividing and their nuclei are in the *interphase* stage of the cell cycle, the chromosomes appear as unorganized, intermingled chromatin fibers. But, in fact, chromosomes occupy individual regions of the cell nucleus, referred to as *territories* (Fig. 1). The arrangement of these territories is usually the same in the nuclei of a given cell type but may be different in the cells of other tissues [1]. Each chromosome contains regions of active chromatin wherein the genes are being expressed (euchromatin) and regions of inactive chromatin where the transcriptional units are silenced or absent (heterochromatin). The active chromatin regions of the different chromosomes tend to be localized in the central portion of the nucleus, referred to as *compartment A*, while inactive regions remain on the periphery and associate with the internal surface of the nuclear membrane (the *nuclear lamina*; see the following) and constitute *compartment B.*

Topological associating domains: Beyond their distinct localization within nuclei, the chromosomes exhibit an amazing level of internal organization that is directly related to the differential expression of genes in the various cell types (Fig. 1). The major architectural features include *topologically associating domains* (TADs) and *functional loops* (nanodomains where most enhancer–promoter associations take place).

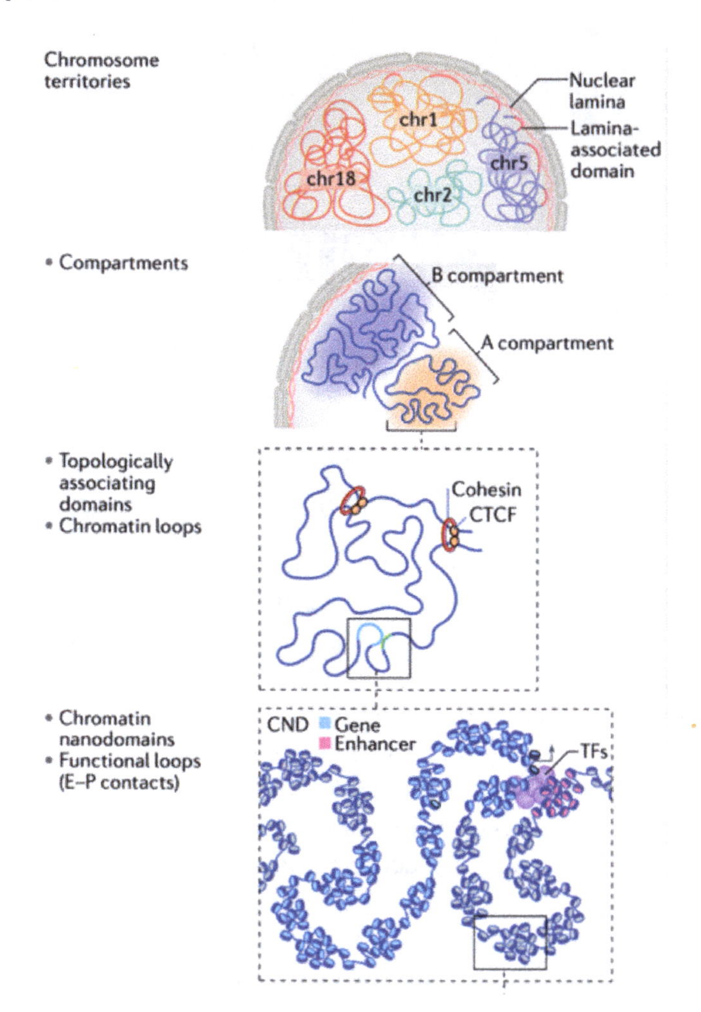

Fig. 1. Different levels of chromosome organizations in mammals. Cohesin is a protein complex that forms rings around different regions of the chromosome that associate via the binding of CTCF proteins; TFs are transcription factors (modified from Jerkovic and Cavalli [2]).

Topologically associating domains represent loops of approximately ten to hundreds of kilobases (a kilobase is 1,000 nucleotides) in length. TADs are formed by *loop extrusion*, a process that is carried out by *cohesin* complexes (see Box 1). In mammals, TAD boundaries are most often

Box 1

TAD formation by loop extrusion mediated by cohesin: Cohesin is a complex of four different proteins (Fig. 2): SMC1, SMC3 (structural maintenance of chromosomes), RAD21 and STAG1/2 (stromal antigen 1). The two SMC proteins form a cassette that can bind adenosine triphosphate (ATP) and hydrolyze it to release energy. Cohesin is found throughout the genome, that it accesses at the binding sites of a loading factor (NIPBL, Nipped-B-like protein). It performs different functions that include sister chromatid cohesion during cell division, DNA repair and transcription regulation (by tethering enhancers to gene promoters). It is also responsible for the formation of TADS by the process of loop extrusion (Fig. 3). Although loop extrusion has not been seen in living nuclei, a number of *in vitro* experiments have demonstrated its occurrence *in vitro*. Using the energy liberated by ATP hydrolysis, cohesin extrudes DNA loops until it meets two sites that are bound by the CTCF protein; usually, the two binding sites are in opposite (convergent) orientation.

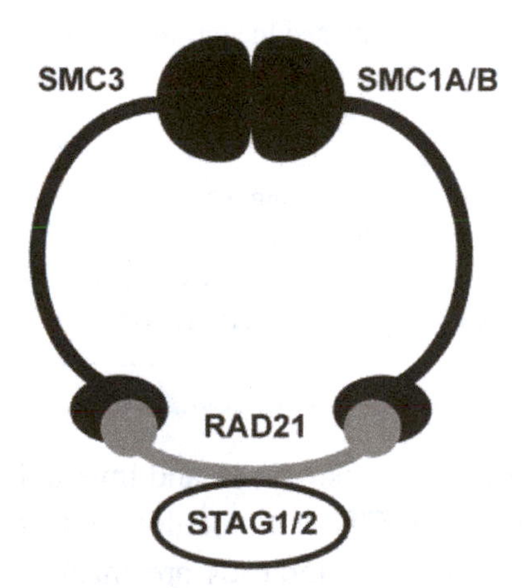

Fig. 2. Diagram illustrating the structure of the cohesin complex. The two ovals at the top of the SMC proteins form the ATPase cassette.

Not surprisingly, given their ubiquitous role in various aspects of chromatin regulation, long non-coding RNAs bind to the CTCF molecules and

(*Continued*)

Box 1 (*Continued*)

cohesin complexes found at TAD boundaries; although their specific functional roles remain to be determined, there is little doubt regarding their regulatory nature.[a]

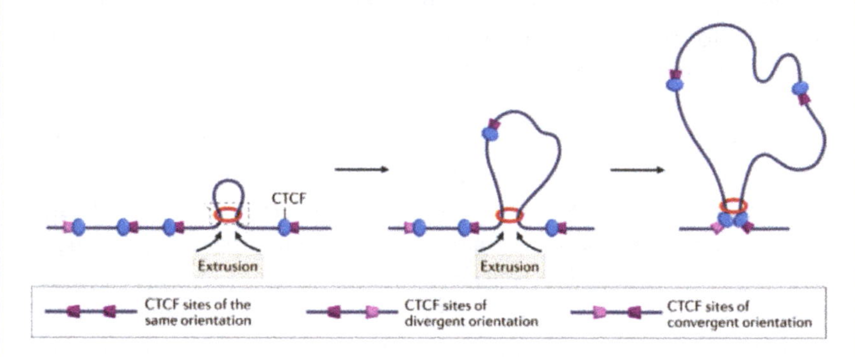

Fig. 3. Diagram illustrating the loop extrusion process by the condensing complex. Extrusion, i.e., loop growth, stops when two CTCF sites, one on each of the two regions forming the loop and positioned in convergent orientation, are brought into binding proximity at the site of the condensin complex (modified from Zhang, *et al.*, 2022[b]).

[a]Yamamoto, T., and N. Saitoh, Non-coding RNAs and chromatin domains. *Current Opinion in Cell Biology*, 2019. **58**: 26–33.
[b]Zhang, Y., *et al.*, The role of chromatin loop extrusion in antibody diversification. *Nature Reviews Immunology*, 2022. doi: 10.1038/s41577-022-00679-3: 1–17.

defined by the presence of cohesin rings and inverted CTCF binding sites that allow the association of CTCF molecules[1] and stop the extrusion process. In other species, TAD boundaries are marked by the presence of other insulator proteins [3]. TADs include subdomains of active and inactive chromatin, with the former containing genes that are coregulated, for example, during differentiation. The contacts between chromatin regions

[1]CTCF (CCCTC-binding factor) is a protein that binds to three regularly spaced repeats of the core DNA sequence CCCTC. Cohesin is a complex of four proteins that form a ring. When brought into contact, CTCF and cohesin molecules bind with each other.

within a TAD, such as those between a gene promoter and its enhancer(s), are much more frequent than they are between TADs. The chromatin of a TAD is subdivided into different nanodomains consisting of clusters of nucleosomes. Histone acetylation, a modification generally associated with active transcriptional units, reduces the number of clusters suggesting that they may interfere with enhancer–promoter interactions.

Association of promoters with enhancers: Promoters (genomic regions where gene transcription is initiated) and enhancers (regions that allow and amplify transcription) physically associate to initiate transcription by forming chromatin loops. These loops are stabilized by CTCF molecules that are present at sites near enhancers and bind with cohesin near gene promoters (Fig. 4).

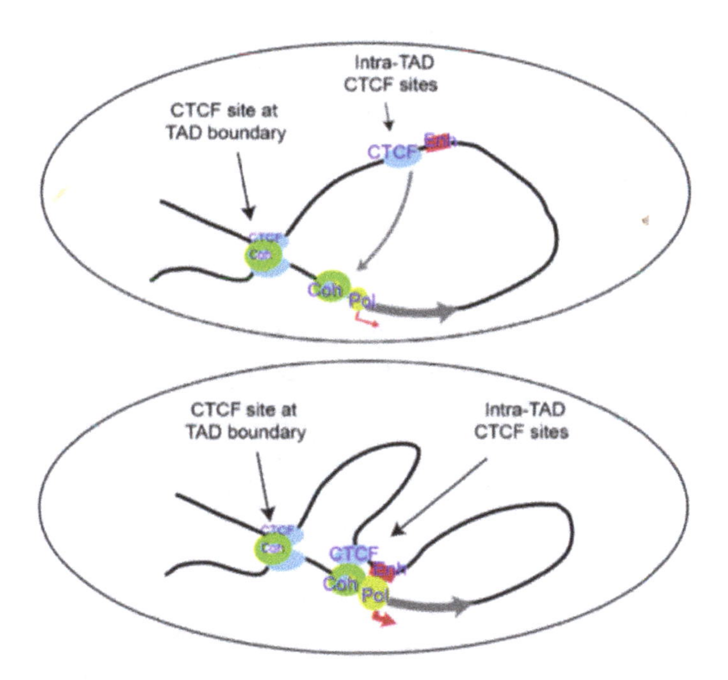

Fig. 4. The association of cohesin (Coh) and CTCF molecules establishes the boundaries of TADs. CTCF collaborates with cohesin to establish and stabilize the association of promoters (indicated by Pol, abbreviation for RNA polymerase) with their enhancers (Enh) within TADs (from Ren *et al.* [4]).

Insulator elements: Insulators play a key role in genome architecture. First discovered in the fruit fly Drosophila, insulators occur in the genome of all eukaryotic organisms. Specific proteins bind to these sequences and are responsible for insulator function. In vertebrates, one of the most common and best-studied insulators is the binding site of the CTCF protein. Insulators appear to affect gene activity by physically blocking the interaction between a gene promoter and its enhancer: the CTCF protein can associate with transcription factors present at promoters and prevent enhancer–promoter contact [5].

B. Functional Subdivisions of the Nucleus

The nuclear envelope: The nucleus is surrounded by a *nuclear envelope* consisting of two lipid bilayers, traversed by a series of openings, the *nuclear pores* [6]. The outer bilayer is continuous with the *endoplasmic reticulum* (an interconnected network of flattened membrane sacs found in all eukaryotic cells where the synthesis, modifications and transport of proteins and some lipids occur). The inner bilayer is associated with the *nuclear lamina*, a mesh of filamentous (*lamins*) and other proteins that maintains the shape of the nucleus and plays a role in the functional organization of chromatin. The nuclear pores are channels made up of many proteins (*nucleoporins*) that allow the regulated transport of molecules and molecular complexes in and out of the nucleus (Fig. 5). A substantial portion of the chromatin within the nucleus, mostly consisting of silent or gene-poor regions of the genome (heterochromatin), is associated with the lamina by means of different proteins that recognize particular DNA sequences. This association is dynamic in that transcriptional units move away from the lamina in those cell types where they are active.

The nucleus does not simply perform the single function of containing the genetic material. It is a highly organized structure made up of numerous subnuclear regions (*nuclear compartments*) that are involved in every aspect of gene expression. In contrast to the many cellular components that are bound by membranes, nuclear compartments are membraneless and consist of localized condensations of DNA, RNA and proteins

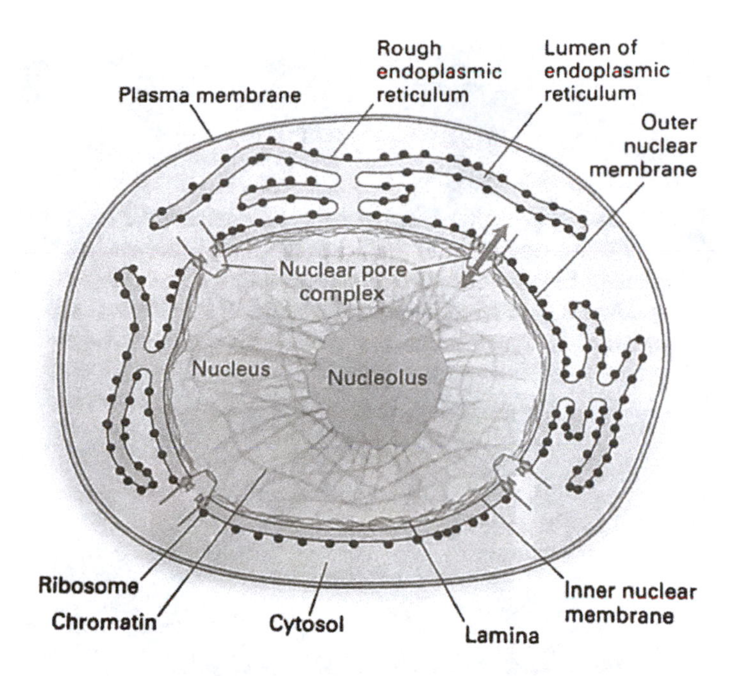

Fig. 5. Diagram of a typical eukaryotic cell showing the relationship of the two-component membranes of the nuclear envelope with the endoplasmic reticulum and the nuclear lamina. The ribosomes are the site of mRNA translation and protein synthesis; the nucleolus is a subnuclear compartment where ribosomes are assembled (from *Molecular Cell Biology*, 9th edn).

(Fig. 6). These condensates are thought to promote molecular interactions by concentrating enzymes, substrates and cofactors or, conversely, to prevent interactions by sequestering enzymes and cofactors. Although specific lncRNAs are known to be involved, the biophysical principles underlying the formation of these regions are still under investigation [7]. The most prominent of the nuclear compartments is the *nucleolus*.

The nucleolus: The *nucleolus* is the site of biogenesis of ribosomes, the cellular particles that carry out the translation of mRNAs into proteins. Nucleoli form on *nucleolus organizer* regions, present on particular chromosomes; these regions consist of clusters of repeated transcription units

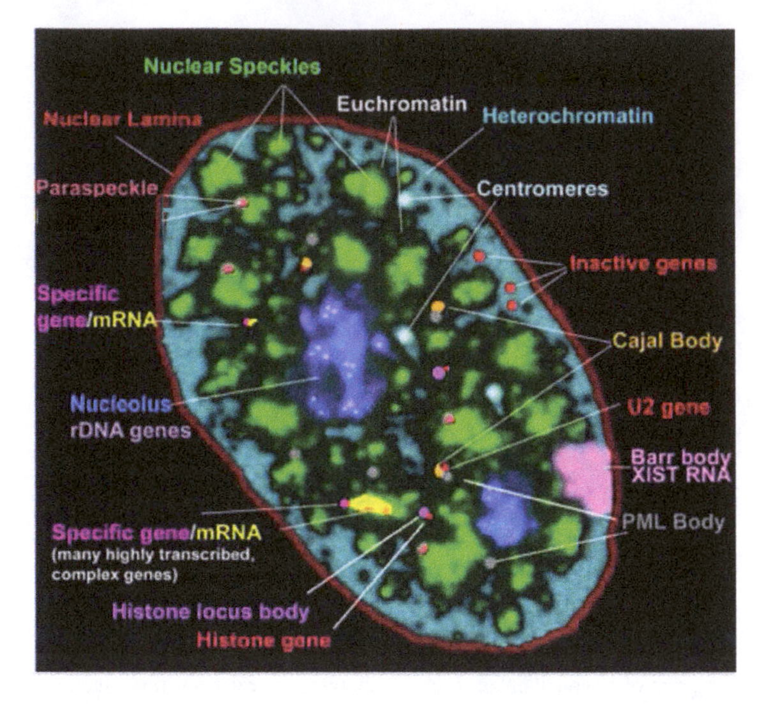

Fig. 6. Composite diagram illustrating the different compartments present in the nucleus (modified from Smith *et al.*, 2020 [7]).

responsible for the synthesis of large transcripts that are then cleaved into the three different types of RNAs found in ribosomes (rRNAs). Ribosomal genes are transcribed by RNA polymerase I (RNAPI). During processing, rRNAs undergo extensive modifications (RNA editing) consisting of ribose methylation and the conversion of uridine into pseudouridine. A fourth RNA and the many ribosomal proteins are transcribed elsewhere in the genome and are targeted to the nucleolus for ribosomal assembly.[2] Not all rRNA genes are active in a given cell type and the relative number

[2] Mature ribosomes of eukaryotic cells consist of a small (40S) and a large (60S) subunit (S stands for Svedberg units that represent the rate of sedimentation during centrifugation based on the mass, density and shape of a molecule or a complex). The small ribosomal subunit includes approximately 20 proteins and an 18S rRNA. The large subunit includes approximately 60 proteins plus 28S, 5.8S rRNAs, as well as a 5S rRNA.

of active genes varies depending on the requirement for protein synthesis during development, in a particular cell type or upon the onset of oncogenesis. Cleavage and editing of the pre-rRNA transcript are dependent on the presence of different *small non-coding nucleolar RNA protein complexes* (snoRNPs) that are thought to form in the nucleoli and in the Cajal bodies (see the following).

In addition to ribosome biogenesis, the nucleolus carries out a number of important nuclear functions that include gene repression, DNA damage repair, responses to stress and telomere maintenance (see Box 2). The silent rRNA gene arrays form a peripheral region of heterochromatin

Box 2

The many functions of the nucleolus: In humans, there are over 4,500 proteins in the nucleolus and only 30% of them are responsible for the synthesis of ribosomes. The other proteins are involved in a number of unrelated functions associated with the regulation of the cell cycle, with DNA damage repair and with responses to stress. Cellular stress responses provide a defense reaction to the damage that environmental factors, such as UV or ionizing radiation, toxic chemicals, free oxygen radicals or sudden temperature changes (heat shock), inflict on cellular molecules.[a]

Over 160 nucleolar proteins are involved in DNA repair. These proteins are sequestered in the nucleolus, to be released when DNA damage occurs. Some of the DNA repair proteins are also involved in monitoring the synthesis of ribosomes. For example, one of the key proteins of the base excision DNA repair pathway (see Chapter 8) recognizes and removes faulty ribosomal RNA transcripts.

In addition to DNA damage, cytological stress can involve changes in protein concentration or can induce structural changes that can affect protein function. Misfolded proteins form aggregates in the nucleolus that can be degraded by the *proteasomes*, protein complexes that contain enzymes that break the bonds between amino acids.

If the cellular stress exceeds the ability of the response mechanisms to intervene, the cells will die, a process referred to as *apoptosis*. This is primarily the responsibility of p53, a protein encoded by a tumor suppressor

(*Continued*)

Box 2 (*Continued*)

gene (see Chapter 11). This factor is normally kept at a low level in cells to control the cell cycle. Its induction at high levels mediated by the nucleolus induces cell cycle arrest and apoptosis.

Finally, the nucleolus is the site of assembly of telomerase, the enzymatic complex that protects the ends of chromosomes from degradation (see Chapter 11).

---

[a]Iarovaia, O.V., Nucleolus: A central hub for nuclear functions. *Trends in Cell Biology*, 2019. **29**(8): 647–659.

with which other genomic gene-poor or silent regions, enriched in repressive histone modifications, such as methylated DNA and methylated lysines of histone H3 (H3K9m3, H3K20me3 and H3K27me3), associate (8). DNA can be damaged by exposure to chemicals, UV or ionizing radiation, resulting in mutations that can be harmful to the cell and to the organism (see Chapter 8). The nucleolus includes a number of factors involved in DNA repair. Cellular stress can be caused by exposure to extreme temperatures, toxins or mechanical damage. When such conditions are sensed by particular nucleolar proteins, a series of interactions occur resulting in the activation of such factors as p53, a tumor suppressor protein that stops the cell from dividing and programs it for *apoptosis* (cell death), which is avoided if the damage can be repaired. Note that these proteins are also activated if the DNA damage is irreparable. Finally, the nucleolus participates in telomere maintenance. Telomeres are segments of repetitive DNA and special proteins that are present at the end of chromosomes to protect them from degradation. The repetitive DNA is added by *telomerase*, an enzyme complex that contains an RNA template used to synthesize the DNA repeats (see Chapter 11). Telomerase is assembled in the nucleolus [8].

Nuclear bodies: Nuclear bodies are regions of molecular condensation that include proteins and nucleic acids. They participate in a variety of cellular processes, such as RNA metabolism, ribosome biogenesis and DNA repair. The concentration of particular molecules in these bodies can accelerate reactions, e.g., by concentrating enzymes with a particular set of substrates; they could also sequester molecules and prevent them from acting ubiquitously in the nucleus.

The most prominent of these nuclear compartments are the *speckles.* Many active genes are associated with these entities that number from one to two dozen per nucleus. Speckles are hubs of concentrated factors involved in the process of splicing primary transcripts into mature mRNAs and in their export to the cytoplasm [9]. The primary transcripts of protein-coding genes and of many lncRNAs contain alternating segments that should be retained in the mature transcript (exons) and segments that should be removed (introns). Most introns are removed as the nascent transcript is being synthesized (see Chapter 3) but in a number of cases, as transcription is terminated the resulting RNAs still contain some introns. These unspliced RNAs will either be degraded or will be subjected to post-transcriptional splicing in the nuclear speckles and then prepared for export.

Another type of membraneless structure in the nucleus is the *paraspeckles.* They are tethered to multiple sites of genomic DNA and, although many questions remain regarding their formation and their molecular function, evidence has accumulated that they are involved in cellular differentiation during development [10].

Cajal bodies, named after the scientist who discovered them, are usually found close to nucleoli with which they have a close functional relationship [11]. For example, Cajal bodies are involved in the modification of small non-coding RNAs included in *small nucleolar ribonucleoproteins* (snoRNPs) that participate in rRNA processing and in *small nuclear ribonucleoproteins* (snRNPs) that are involved in pre-mRNA splicing. In addition, they may be involved in DNA repair. Some experiments suggest that Cajal bodies are dispensable, although they increase the efficiency of nucleolar processes, such as the assembly of snRNPs and snoRNPs.

Several other nuclear compartments have been identified [12]. *Premyelocytic leukemia* (PML) *bodies* (also known as ND10 bodies) were discovered as nuclear structures that were absent in the cells of premyelocytic leukemia patients. PML bodies appear to be involved in the deposition of the histone H3 variant (H3.3) into active chromatin regions and contain or are associated with a number of histone-modifying enzymes that neither enhance nor repress gene activity. In the nucleus, RNAPII molecules are concentrated in numerous sites referred to as *transcription factories* in which highly transcribed genes are present; these clusters are also enriched in the Mediator transcription factor. A special case of transcription factories is the *histone locus bodies* that form at the site of transcription of the histone gene clusters and where the transcription of these clusters occurs. *GW bodies,* named after one of the proteins that characterize them, contain miRNAs and other factors that cleave mRNAs in order to abrogate gene expression.

References

[1] Fritz, A.J., *et al.*, Cell type specific alterations in interchromosomal networks across the cell cycle. *PLOS Computational Biology*, 2014. **10**(10): Article ID e1003857, 13 pages.

[2] Jerkovic, I., and G. Cavalli, Understanding 3D genome organization by multidisciplinary methods. *Nature Reviews Molecular Cell Biology*, 2021. **22**(8): 511–528.

[3] Szabo, Q., F. Bantignies, and G. Cavalli, Principles of genome folding into topologically associating domains. *Science Advances*, 2019. **5**(4): Article ID eaaw1668, 12 pages.

[4] Ren, G., *et al.*, CTCF-mediated enhancer-promoter interaction is a critical regulator of cell-to-cell variation of gene expression. *Molecular Cell*, 2017. **67**(6): 1049–1058.

[5] Melnikova, L.S., P.G. Georgiev, and A.K. Golovnin, The functions and mechanisms of insulators in the genomes of higher eukaryotes. *Acta Naturae*, 2020. **12**(4): 15–33.

[6] De Magistris, P., and W. Antonin, The dynamic nature of the nuclear envelope. *Current Biology*, 2018. **28**(8): R487–R497.

[7] Smith, K.P., L.L. Hall, and J.B. Lawrence, Nuclear hubs built on RNAs and clustered organization of the genome. *Current Opinion in Cell Biology*, 2020. **64**: 67–76.

[8] Iarovaia, O.V., *et al.*, Nucleolus: A central hub for nuclear functions. *Trends in Biology*, 2019. **29**(8): 637–659.

[9] Gordon J.M., D.V. Phizicky, and K.M. Neugebauer, Nuclear mechanisms of gene expression control: Pre-mRNA splicing as a life or death decision. *Current Opinion in Genetics and Development*, 2021. **67**: 67–76.

[10] Grosch, M., *et al.*, Chromatin-associated membraneless organelles in regulation of cellular differentiation. *Stem Cell Reports*, 2020. **15**(6): 1220–1232.

[11] Trinkle-Mulcahy, L., and J.E. Sleeman, The Cajal body and the nucleolus: "In a relationship" or "it's complicated"? *RNA Biology*, 2017. **14**(6): 739–751.

[12] Corpet, A., *et al.*, PML nuclear bodies and chromatin dynamics: Catch me if you can! *Nucleic Acids Research*, 2020. **48**(21): 11890–11912.

CHAPTER 6

How Does Transcription Occur in Chromatin?

A. Initiation and Maintenance of Transcription

In its ground state, chromatin is physically compacted and is not accessible to the many factors involved in transcription. The first step for the activation of genes and other transcription units is the binding of factors, often referred to as *pioneer factors*, to their specific DNA binding sites in condensed chromatin [1]. These binding sites are transiently exposed because the association of DNA with the histone octamers is dynamic and some stretches of DNA that are normally tightly wrapped around the octamer are transiently exposed. Since gene activity is different in the various tissues during development and in adult organisms, pioneer factors are cell lineage-specific. In many cases, the pioneer factors associate with coactivators that are present in particular cell types to increase their recognition of specific DNA-binding sites. Once bound, pioneer factors recruit chromatin-remodeling and histone-modifying complexes that open up the chromatin and make the DNA accessible to the transcription machinery (Fig. 1). Another aspect of decondensing chromatin in order to facilitate transcription is the general removal of histone H1 by specific chaperones [3]. In order to carry out transcription, the RNA polymerase complex forces the looping of nucleosome-bound DNA away from the histone octamer. It is assisted in this process by remodeling complexes and histone chaperones [4].

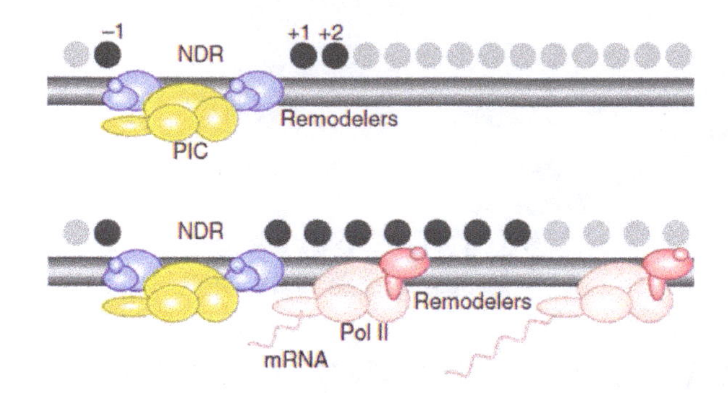

Fig. 1. NDRs are nucleosome-depleted regions that occur as a result of transcription factor and remodeling complex binding. The gray circles indicate nucleosomes; remodeling complexes place nucleosomes at fixed positions (black circles) flanking the NDRs. Downstream of the transcription start sites, the position of nucleosomes depends on the rate of transcription (modified from Struhl and Segal [2]).

The role of remodeling complexes: As previously discussed (Chapter 4), remodeling complexes can reposition nucleosomes along the DNA or completely evict them. In general, promoters and enhancers are depleted of nucleosomes. In addition to their role in transcription initiation, remodeling complexes facilitate the elongation step of the transcription process. As the polymerase complex proceeds, the DNA is transiently unwound; the formation and maintenance of this *transcription bubble* involve the disassembly of nucleosomes in front of the elongating polymerase and their reassembly behind it. This particular feature of transcription is carried out with the collaboration of broadly conserved histone chaperones such as the *facilitate chromatin transcription* (FACT) factor [5].

The role of covalent histone modifications: These modifications, added to histones after their synthesis by a variety of enzyme complexes, are a major regulatory mechanism for gene function. The more widespread and therefore best studied modifications open up the chromatin and assist in the process of transcription; as expected, other modifications are associated with the repression of gene function (see Chapter 4). The acetylation of some lysines (H3K9ac, H3K27ac and H4K16ac) leads to a more open chromatin structure that allows the binding of transcription factors [6]. For example, the acetylation of lysine 16 on histone H4 disrupts the association of neighboring nucleosomes that is a characteristic of condensed and transcriptionally inactive chromatin (see Box 1). Some

Box 1

The role of H4K16 ac in loosening chromatin: The acetylation of lysine 16 on histone H4 is the responsibility of a histone acetyltransferase, first discovered in *Drosophila* and in humans.[a,b] This enzyme is part of a protein complex that is present at numerous active gene promoters and along some transcription units in most organisms. In condensed chromatin, neighboring nucleosomes are associated by the interaction of a basic segment of the tail of histone H4 of one nucleosome with an acidic patch formed by an H2a/H2B histone dimer on the next nucleosome (Fig. 2). The presence of H4K16ac weakens this interaction and thereby facilitates the binding of transcription factors as well as the progression of the transcription complex.[c]

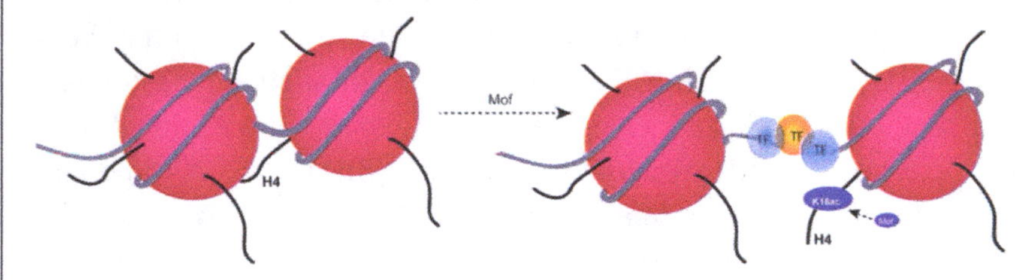

Fig. 2. Loosening of chromatin compaction by acetylation of histone H4 tail. Left: The H4 tail of one nucleosome associates with the acidic patch of the neighboring nucleosome; right: acetylation of the lysine 16 on the tail of H4 reduces the positive charge of the tail and diminishes the interaction with the neighboring nucleosome allowing the association of transcription factors with the uncovered DNA sequences (modified from Pradepa, 2016[d]).

[a]Hilfiker, A., *et al.*, mof, a putative acetyl transferase gene related to the Tip60 and MOZ human genes and to the SAS genes of yeast, is required for dosage compensation in *Drosophila. The EMBO Journal*, 1997. **16**(8): 2054–2060.

[b]Neal, K.C., *et al.*, A new human member of the MYST family of histone acetyl transferases with high sequence similarity to *Drosophila* MOF. *Biochimica et Biophsica Acta*, 2000. **1490**(1–2): 170–174.

[c]Luger, K., *et al.*, Crystal structure of the nucleosome core particle at 2.8 A resolution. *Nature*, 1997. **389**(6648): 251–260.

[d]Pradeepa, M.M. Causal role of histone acetylations in enhancer function. *Transcription*, 2017. **8**(1): 40–47.

acetylated lysines actually serve as targets for specific factors: they are recognized by a particular protein module, the *bromodomain*, found in a large number of chromatin-modifying enzymes such as histone acetyltransferases and methyltransferases, remodeling complex subunits and transcription factors [7].

Lysine 4 of histone H3 trimethylation (H3K4me3) and lysine 27 of histone H3 acetylation, in the promoter region, contribute to gene activation, and methylated lysines 36 or 79 of histone H3 (H3K36me and H3K79me) mark the body of actively transcribed genes; the methylation of H3K79 requires the previous ubiquitination of histone H2b at lysine 120 (H2BK120ub). Other methylated lysines (H3K9me, H3K27me and H4K20me),[1] present throughout the domains of transcriptional units, effect or maintain repression (see the following). In some genes, H3K4me3 (as well as H3K27ac) are present throughout the transcriptional unit, resulting in an increase in the number of transcriptional rounds and, therefore, in the total output of the gene [8]. Another histone methylation, H3K36me, plays multiple roles in gene regulation. Throughout the length of genes, the presence of this modification prevents the inappropriate initiation of transcription at internal sites (rather than at the promoter). In addition, H3K36me participates in DNA damage repair (Chapter 8) and the regulation of pre-mRNA splicing [9].

Enhancers and promoters: Promoters have been defined as regulatory regions where transcription is initiated, and enhancers as regulatory regions that allow and amplify this transcription. Enhancer activity is cell-type specific and often occurs in response to environmental signals. Histone modifications most frequently found at active enhancers are monomethylated lysine 4 and acetylated lysine 27 on histone H3 (H3K4me1 and H3K27ac). Inactive enhancers are marked by H3K27me3. Although promoters and enhancers have been considered to be distinct regulatory elements, it is important to note that they share several structural

[1] Note that some of these modifications, H3K27me3 and H3K9me3, define heterochromatic regions of the genome; H3K9me3 is recognized by molecules of *heterochromatin protein 1* (HP1) that associate with one another (self-oligomerize) to insure the compaction of chromatin.

and operational similarities, based on the observations that some promoters have enhancer activity and that many enhancers are, themselves, the site of RNA transcription. As in the case of gene transcription, the production of *enhancer RNA* (eRNA) involves a promoter region, traditional transcription factors and RNAPII. Most eRNAs are relatively short and are capped but not polyadenylated and, therefore, are relatively unstable. These RNAs are implicated in a variety of regulatory functions, from stabilizing enhancer–promoter contacts to maintaining an open chromatin structure by associating with transcriptional regulators; some of them have been shown to perform the latter functions in trans, i.e., at remote genomic sites [10]. Surprisingly, some promoters appear to work as enhancers in that they can influence the activity of other, distant promoters usually associated with housekeeping genes (genes required for all basic cellular functions and, therefore, active in all cells) or stress response genes [11].

As previously mentioned, the nucleosomes at active promoters are marked by histone H3 lysine 4 trimethylation (H3K4me3) while silentpromoters exhibit H3K27me3. A number of promoters, especially in differentiating cells, are characterized by the presence of both types of histone marks. These so-called *bivalent promoters* are poised for rapid activation. Poised enhancers exist as well and are identifiable by the simultaneous presence of H3K4me1 and H3K27me3 [12].

The Trithorax Group (TrxG) proteins: This large group of evolutionarily conserved proteins, named after the first member that was identified in the fruit fly *Drosophila*, form different complexes that promote active transcription. Some of these complexes constitute the methyl transferases that are responsible for the trimethylation of H3K4 at promoters and its monomethylation at enhancers; others assemble into the remodeling complexes of the SWI/SNF family that reposition nucleosomes to make chromatin more accessible [13].

RNA polymerase II modifications: RNAPII and the other RNA polymerases consist of multiprotein complexes. To initiate transcription, some of the amino acids of the terminal end of the largest subunit of RNAPII must be phosphorylated, i.e., modified by kinases, enzymes that add phosphate groups. Furthermore, to overcome the pausing that occurs following the transcription of the first few nucleotides (see Chapter 3), additional amino acids of the same terminal end must be phosphorylated. Release from pausing also involves the action of kinases that phosphorylate and inactivate the pausing factors DSIF and NELF [14]. The RNAPII is now free to proceed with productive transcription elongation. Its ability to avoid premature termination and dissociation from the DNA template and to reach the end of the transcription unit is termed *processivity* and the speed with which it adds nucleotides to the nascent RNA molecule is the *elongation rate*. This latter parameter is important in regulating the rate of synthesis of gene products [15].

The role of histone variants: The incorporation of histone variants into nucleosomes by specific chaperones alters the organization of chromatin and is thought to contribute to all DNA-associated functions, including transcription. These variants are primarily isoforms of histones H3 and H2A. The H3 variant H3.3 is deposited in recycling nucleosomes of active chromatin including enhancers, promoters and gene bodies. Although it is thought to destabilize the folding of chromatin favoring a more open structure (enrichment of H3.3 is correlated with loss of histone H1), the absence of this variant does not seem to affect the general level of transcription [16], suggesting that there may be compensatory mechanisms based on the incorporation of other variants. Another histone variant, H2A.Z, present in all eukaryotes, accumulates in the promoter region; surprisingly, its role in the regulation of gene expression is different in different organisms: in some, it facilitates transcription elongation, while in others, it associates with silenced genes. H2A.Z is also found in enhancers but its effect on enhancer function is not clear [17].

B. Repression of Transcription

Just as important as the activation of genes that are required to achieve specific aspects of cellular differentiation and function is the repression of genes that are no longer required or that are irrelevant to these processes. Gene silencing is the responsibility of complexes that erase some of the histone marks characteristic of gene activation and replace them with repressive marks.

The Polycomb Group (PcG) proteins: Named, once again, after the first gene discovered in *Drosophila*, these proteins form two types of *Polycomb repressive complexes*: PRC1 that ubiquitinates lysine 119 of histone H2A (H2AK119ub), a modification known to facilitate chromatin compaction, and PRC2 that catalyzes the methylation of histone H3 lysine 27 (H3K27me1, me2 or me3, the latter being the canonical repressive mark). The localization of both complexes occurs at genomic sites termed *Polycomb responsive elements* (PREs) and requires different binding factors. In some cases, the presence of PRC2 and H3K27me3 is required for the binding of PRC1 and the ubiquitination of H2AK119; in other cases, PRC1 complexes, albeit with different subunit compositions, bind directly to PREs where they attract PRC2 complexes via a subunit that recognizes the ubiquitin moiety [13,18].

The role of DNA methylation: In many, but not all organisms, DNA methylation is an important factor that influences transcription by modulating the binding of transcription factors. As mentioned in Chapter 4, this modification occurs mainly on the cytosine of cytosine–guanine (CpG) dinucleotides. CpGs occur throughout the genome but are particularly enriched in short regions of around 500 to 2,000 nucleotides that occur in or near gene promoters. Generally, these regions referred to as *CpG islands* lack methylation if the genes with which they are associated are active, allowing the binding of transcription factors to the promoter for transcription activation. Unmethylated CpG islands can also be associated with active enhancers and non-coding RNA transcriptional units. Hypermethylated CpG islands have been traditionally correlated with gene repression. Several proteins exist that bind to methylated DNA and interact with histone remodelers and modifiers that lead to gene silencing. As is the case with most biological generalizations, exceptions to the stated roles of DNA methylation have been observed: some methylated CpG islands allow the binding of transcription factors to promoters and, therefore, gene activation [19]. Furthermore, experimentally induced DNA methylation has recently been shown to silence genes that do not have a CpG island associated with their promoter [20]. Another area of the genome where DNA methylation occurs is along the body of genes that are actively transcribing. This modification is thought to facilitate the process of transcription elongation and to prevent the ectopic initiation of transcription at internal sites within genes.

In mammals, the DNA methylation that was present in the genomes of the egg and sperm is erased during very early embryogenesis; it is soon re-established according to the transcriptional programs necessary for cellular differentiation and development. An exception is the methylation of imprinted genes (see the following).

C. Patterns of Individual Gene Expression

Transcriptional bursting: Gene transcription normally occurs in bursts leading to the observation that at any given time, a gene may be active in some of the cells of a particular tissue and inactive in others. While this statement refers to both copies of a gene in diploid organisms, there are instances when, in some cells of a given tissue, only one of the two alleles is transcribed, while in other cells, the other allele is activated [21]. Another type of monoallelic expression occurs in mammals and some other organisms and is referred to as *genomic imprinting*.

Monoallelic expression and genomic imprinting: This phenomenon was discovered when experimentally derived mouse eggs containing two copies of the maternal or paternal genomes (rather than one of each, as a result of normal fertilization) produced inviable embryos. These observations led to the conclusion that the parental genomes are not equivalent and that some genes are expressed according to their parental origin in a developing embryo [22]. Since the alleles of imprinted genes that are inactive in an individual of a particular sex will be active when they are transmitted to progeny of the opposite sex (or vice versa, i.e., active alleles in one individual will become inactive in opposite sex progeny), imprinting is not the result of genetic mutations; rather, it is an epigenetic phenomenon. Imprinted genes usually occur in clusters that are regulated (silenced) by *imprinting control regions* (ICRs). If the ICR sequence is methylated, the imprinted alleles that it regulates will be expressed; if it is unmethylated, a regulatory lncRNA is synthesized and prevents the activation of the imprinted alleles [24]. During the formation of gametes, the ICRs that regulate genes to be expressed only from the paternal alleles are methylated during spermatogenesis and those that regulate genes to be expressed from the maternal alleles are methylated during oogenesis. These modifications are retained during the reprogramming of DNA methylation that occurs during embryonic development discussed above (Fig. 3).

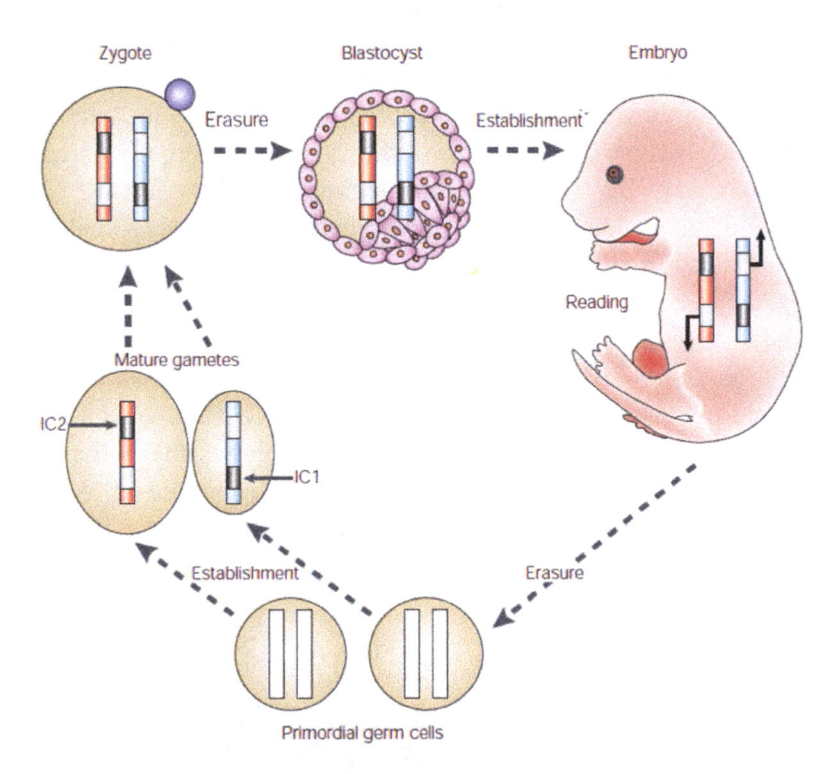

Fig. 3. Diagram of the genomic imprinting cycle. Chromosomes inherited from the male parent are marked in blue and those from the female parent in red. IC1 and IC2 represent two imprinting control regions, shaded gray if they are methylated and inactive and white if they are active. The smaller gamete represents a sperm, the larger an egg. Because the patterns of DNA methylation of the egg and sperm are different, they are erased soon after fertilization and re-established during embryonic development, with the exception of imprinted IC regions (and the imprinted alleles that they control) which remain methylated. As germ cell precursors are formed in the developing gonads, the methylation of their genomes including imprinted genes is erased. During the early stages of gametogenesis, *de novo* methylation occurs leading to the levels of methylation found in mature gametes and including the methylation of imprinting control regions according to the developing embryo's sex (from Reik and Walter [23]).

D. Transcription of Gene Clusters

In order to achieve cellular differentiation and the maintenance of cellular function, groups of cognate genes must be activated simultaneously. This can be achieved by the presence of common binding sequences for transcription factors in the regulatory regions of these genes. Alternatively, the genes may be grouped into clusters that can be regulated as a unit. Such clusters may have arisen during the course of evolution by the duplication of an original gene followed by the diversification in function of the gene and its copies. An example of this type of transcriptional regulation is provided by the α-globin and β-globin genes of vertebrates.[2] Other examples of gene clusters that are coregulated are the ribosomal RNA genes and the histone genes. The β-globin gene cluster involves genes that are required during embryonic, fetal or adult stages (Fig. 4). The sequential activation of these different genes during development and adult life is achieved by their individual interaction with an enhancer region referred to as the *locus control region* (LCR). The sequential activation of β-globin genes during development requires the silencing of those genes that should remain silent. This is achieved by the DNA methylation of these genes and its recruitment of silencing histone-modifying complexes [25].

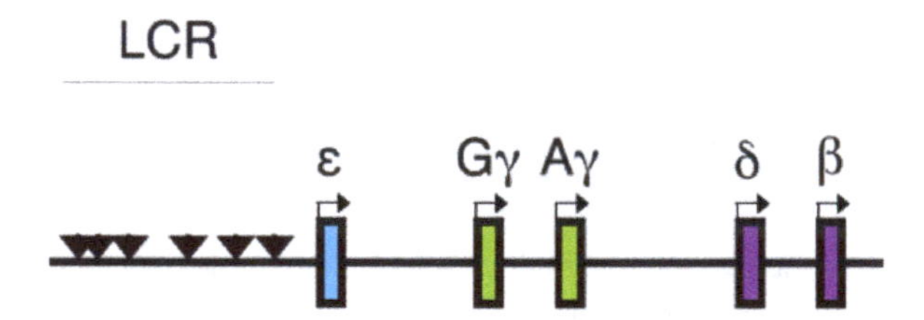

Fig. 4. The human β-globin locus consists of a gene for embryonic (ε), two genes for fetal ($^{G}\gamma$, $^{A}\gamma$) and two genes for adult hemoglobins (δ, β). The LCR is a series of enhancer DNA sequences.

---

[2]The oxygen carrier in the vertebrate circulatory system is hemoglobin that consists of α-globin and β-globin chains. During the course of evolution, an ancestral gene was duplicated, and further duplication of each copy gave rise to two separate gene clusters that encode the α-globin and β-globin chains.

E. Transcription Regulation of Whole Chromosomes

One of the best-studied and most relevant examples is the transcriptional regulation of entire sex chromosomes. In many animal species, the sex of individuals is determined by the presence of sex chromosomes. In mammals and in fruit flies, the sex chromosomes are represented by an X and a Y in males and two X chromosomes in females. In mammals, the Y chromosome is male-determining; in fruit flies, it is the presence of one versus two X chromosomes that determines male differentiation.[3] In both instances, a problem arises with respect to the doses of genes present on the X chromosome in the two sexes. Although some of these genes' functions are sex-related, many are responsible for basic steps in cell differentiation and function that are equally important for the two sexes; their presence in two doses in females and a single dose in males would lead to a disparity in these basic gene products. To avoid this problem, epigenetic mechanisms compensate for this difference in dosage and equalize the level of transcription of X-linked genes between the sexes [26]. In Drosophila, where the first mechanisms for X chromosome dosage compensation were discovered, the level of transcription of most X-linked genes is rendered equal in males and females by the global acetylation of histone H4 at lysine 16 (H4K16ac) in the X chromosome chromatin of males. The acetyltransferase responsible is present in a regulatory complex (the male-specific lethal (MSL) complex) that assembles on a long non-coding RNA and contains a helicase that unwinds RNA:DNA duplexes. The MSL complex assembles only in males and recognizes specific DNA sequences along the X chromosome from where it spreads to active genes and enhances their level of activity by facilitating transcription elongation.

In humans and all placental mammals, one of the two X chromosomes becomes condensed (heterochromatic) in early female embryos resulting

---

[3] In other organisms, the only sex chromosome in males is an X. In still others, males have two identical sex chromosomes (Z chromosomes) and females have one Z and another sex chromosome that is different from Z and is called the W chromosome.

in the simultaneous repression of most genes.[4] A gene present on the X chromosome leads to twice the amount of gene product in females than in males; this is sufficient to activate the *Xist* (X inactive transcript) gene on one of the two X chromosomes. The lncRNA produced spreads along the X chromosome and recruits factors and enzymatic complexes that effect the epigenetic modifications responsible for chromatin silencing: DNA methylation, histone H4 deacetylation, reduction of H3K4me and increase in H3K9me3 and H4K20me3. In addition, the histone H2A variant *macroH2A* is enriched on the inactive X chromosome. In the early embryo, which X chromosome (maternal or paternal) becomes inactivated is random but, once the choice is made in a particular cell, all of its descendants will inactivate the same X. A gene is present very close to the *Xist* gene that is transcribed in the opposite direction. This gene (*Tsix*, the name is the reverse of *Xist*) is induced on the active X chromosome, and its transcripts interfere with the *Xist* transcripts, thereby ensuring that this chromosome remains active. The signals and factors that turn on the *Tsix* gene are not fully understood.

References

[1] Zaret, K.S., Pioneer transcription factors initiating gene network changes. *Annual Reviews Genetics*, 2020. **54**: 367–385.

[2] Struhl, K., and E. Segal, Determinants of nucleosome positioning. *Nature Structural and Molecular Biology*, 2013. **20**(3): 267–273.

[3] Zhang, Q., *et al.*, Eviction of linker histone H1 by NAP-family histone chaperones enhances activated transcription. *Epigenetics and Chromatin*, 2015. Article ID PMC4558729, 17 pages.

[4] Kujirai, T., and H. Kurumizaka, Transcription through the nucleosome. *Current Opinion in Structural Biology*, 2020. **61**: 42–49.

[5] Formosa, T., and F. Winston, The role of FCT in managing chromatin: Disruption, assembly, or repair? *Nucleic Acids Research*, 2020. **48**(21): 11929–11941.

---

[4]In marsupials, it is the paternal X that becomes inactivated in females while in monotremes (egg-laying mammals), which possess several X and Y chromosomes, dosage compensation occurs on a gene-by-gene basis.

[6] Chen, Y.-J.C., *et al.*, Now open: Evolving insights to the role of lysine acetylation in chromatin organization and function. *Molecular Cell*, 2022. **82**(4): 716–727.
[7] Lloyd, J.T., and K.C. Glass, Biological function and histone recognition of family IV bromodomain-containing proteins. *Journal of Cell Physiology*, 2018. **233**(3): 1877–1886.
[8] Beacon, T.H., *et al.*, The dynamic broad epigenetic (H3K4me3, H3K27ac) domain as a nark of essential genes. *Clinical Epigenetics*, 2021. **13**(1): 1–17.
[9] McDaniel, S.L., and B.D. Strahl, Shaping the cellular landscape with Set2/SETD2 methylation. *Cellular and Molecular Life Sciences*, 2017. **74**(18): 3317–3334.
[10] Sartorelli, V., and S.M. Lauberth, Enhancers RNAs are an important regulatory layer of the genome. *Nature Structural and Molecular Biology*, 2020. **27**(6): 521–528.
[11] Dao, L.T.M., and S. Spicuglia, Transcriptional regulation by promoters with enhancer function. *Transcription*, 2018. **9**(5): 307–314.
[12] Blanco, E., *et al.*, The bivalent genome: Characterization, structure, and regulation. *Trends in Genetics*, 2019. **36**(2): 118–131.
[13] Kuroda, M.I., *et al.*, Dynamic competition of Polycomb and Trithorax in transcriptional programming. *Annual Reviews of Biochemistry*, 2020. **89**: 235–253.
[14] Dollinger, R., and D.S. Gilmour, Regulation of promoter proximal pausing of RNA polymerase II in metazoans. *Journal of Molecular Biology*, 2021. **433**(14): 1–21.
[15] Muniz, L., E. Nicolas, and D. Trouche, RNA polymerase II speed: A key player in controlling and adapting transcriptome composition. *The EMBO Journal*, 2021. **40**(15): 1–21.
[16] Shi, L., H. Weng, and X. Shi, The histone variant H3.3 in transcriptional regulation and human disease. *Journal of Molecular Biology*, 2017. **429**(13): 1934–1945.
[17] Scacchetti, A., and P.E. Becker, Variation on a theme: Evolutionary strategies for H2A.Z exchange by SWR1-type remodelers. *Current Opinion in Cell Biology*, 2021. **70**: 1–9.
[18] Cohen, I., C. Bar, and E. Ezhkova, Activity of PRC1 and histone H2AK119 monoubiquitination: Revising popular misconceptions. *Bioessays*, 2020. **42**(5): 1–17.

[19] Angeloni, A., and O. Bogdanovic, Sequence determinants, function, and evolution of CpG islands. *Biochemical Society Transactions*, 2021. **49**(3): 1109–1119.

[20] Nunez, J.K., *et al.*, Genome-wide programmable transcriptional memory by CRISPR-based epigenome editing. *Cell*, 2021. **184**(9): 2503–2519.

[21] Eckersley-Maslin, M.A., and D.L. Spector, Random monoallelic expression one allele at a time. *Trends in Genetics*, 2014. **30**(6): 237–244.

[22] Tucci, V., *et al.*, Genomic imprinting and physiological processes in mammals. *Cell*, 2019. **176**(5): 952–965.

[23] Reik, W., and J. Walter, Genomic imprinting, parental influence on the genome. *Nature Reviews Genetics*, 2001. **2**(1): 21–32.

[24] MacDonald, W., and M.R.W. Mann, Long noncoding RNA functionality in imprinted domain regulation. *PLOS Genetics*, 2020. **16**(8): 1–22.

[25] Iarovaia, O.V., *et al.*, Genetic and epigenetic mechanisms of β-globin gene switching. *Biochemistry (Moscow)*, 2018. **83**(4): 381–392.

[26] Lucchesi, J.C., Transcriptional modulation of entire chromosomes: Dosage compensation. *Journal of Genetics*, 2018. **97**(2): 357–364.

CHAPTER 7

How Does DNA Replication Occur in Chromatin?

Since DNA is the universal carrier of genetic information, it stands to reason that unicellular organisms as well as the individual cells of multicellular organisms must be able to faithfully replicate their DNA content prior to the production of daughter organisms or daughter cells. DNA replication is one of the many steps in chromatin assembly necessary to reconstitute the parental genetic material transmitted to daughter cells; just as in the case of transcription, the replication process, which precedes cell division, involves the participation of dedicated regulatory factors and enzymatic complexes. DNA replication occurs during the *S phase* of the cell cycle (Fig. 1).

A. DNA Replication

Replication initiates at multiple points (*origins*) along the DNA molecule of each chromosome. Initiation consists of three steps: *origin recognition* by a *complex* (ORC), recruitment of additional factors to transform the ORC into a *pre-replication complex* (pre-RC) and the activation of DNA synthesis by the conversion of the pre-RC into a *pre-initiation complex* (pre-IC); the latter contains a *helicase* enzyme that can unwind the DNA and a DNA polymerase that will assemble the new nucleotide chain. Thousands of pre-RCs are assembled during the G1 phase of the cell cycle and are converted into pre-ICs during the G1 to S phase transition.

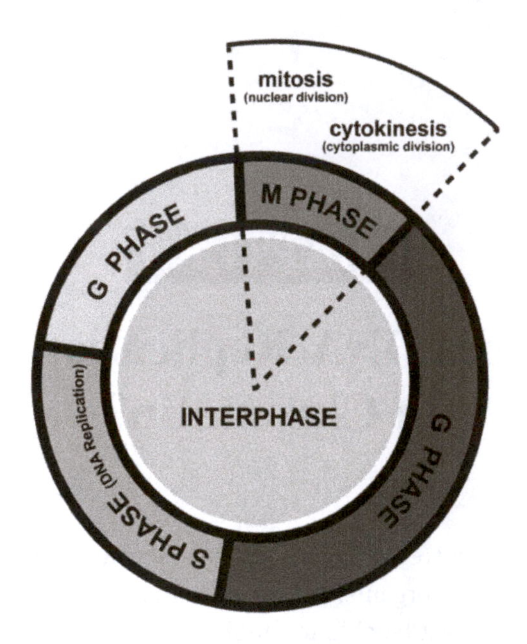

Fig. 1. Diagram of the different stages of the cell cycle. Following the M (mitosis) phase, the daughter cells enter the G_1 (gap 1) phase. DNA replication occurs during the S (synthesis) phase and is followed by a second gap phase (G_2) during which the cell prepares for the next mitotic division.

Unwinding the DNA requires the disassembly and removal of nucleosomes. Activation of the helicase and recruitment of additional factors transform the pre-ICs into *replisomes* and leads to the initiation of replication and the formation of two replication forks that move in opposite directions (Fig. 2). Only a number of the origins that are present in the genome are activated during any S phase.

Clearly, replication of a particular DNA segment and transcription of the genes within that segment cannot occur simultaneously. In order to coordinate their specific gene expression program with the need to undergo DNA replication, different cell types activate different pre-RCs located on different origins of replication to initiate the replication of their genome at a time when the genes in those regions are not active.

A number of epigenetic modifications are associated with the initiation of DNA replication. Although they do not have a specific DNA sequence, origins are characterized by flanking stretches of adenine- and

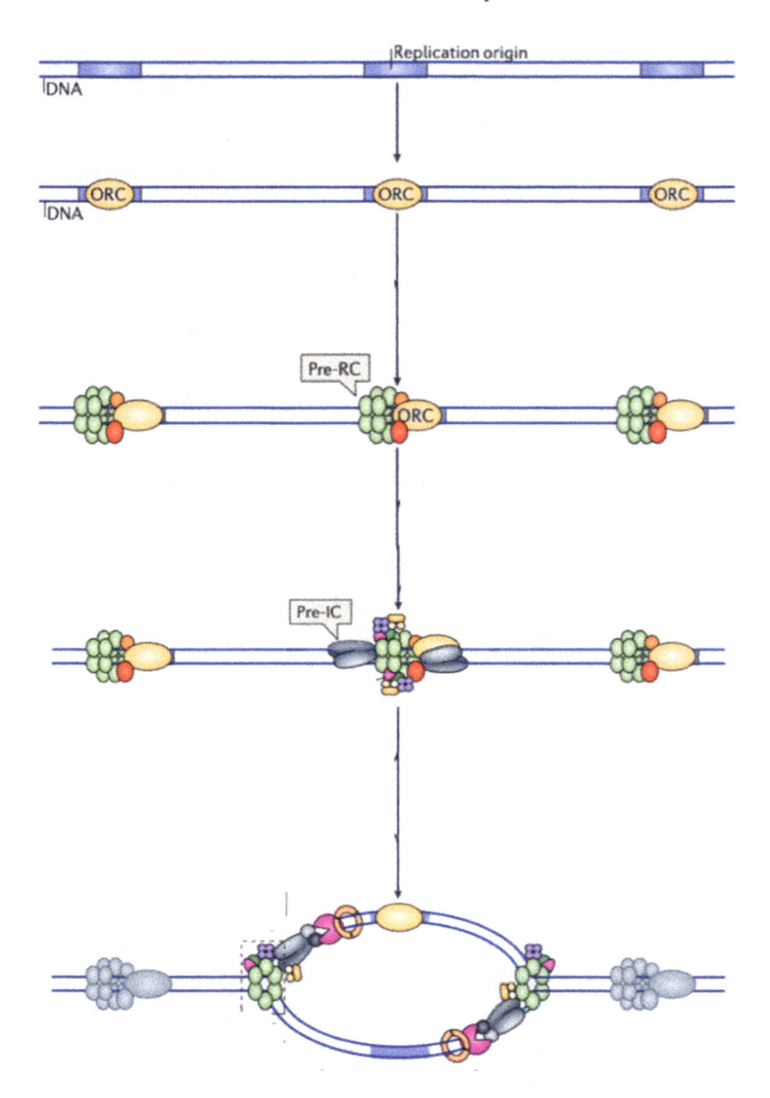

Fig. 2. The different steps of DNA replication consist of recognition of the various origins of replication by a protein complex (ORC), the recruitment of a pre-replication complex (Pre-RC) and its transformation at some but not all ORCs into an active replication complex, the replisome. Clearly, the presence of two diverging replication forks requires the formation of two active complexes at any particular ORC (modified from Fragkos *et al.* [1]).

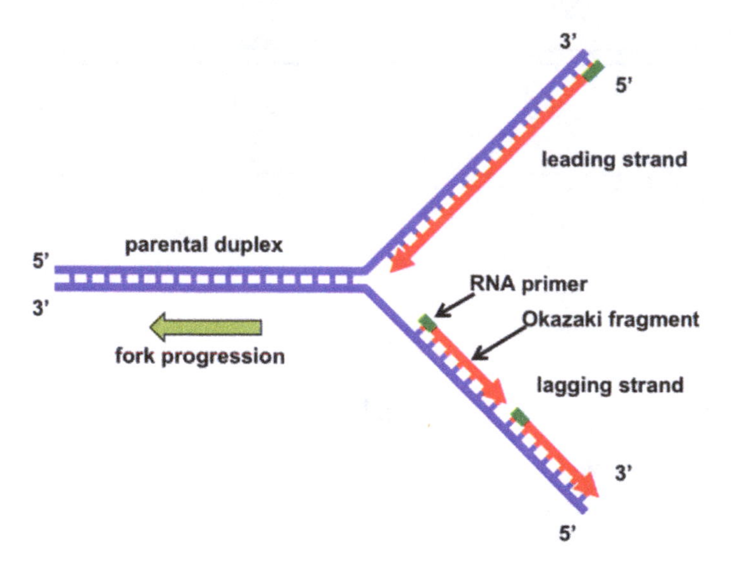

Fig. 3. Diagram of a replication fork illustrating that the old DNA strand that is oriented in a 5′ to 3′ direction (leading strand) can be copied continuously by the DNA polymerase since it adds nucleotides at the 3′ end of the DNA copy that it is synthesizing. The old 3′ to 5′ strand (lagging strand) has to be copied discontinuously: short RNA primers are synthesized that can be extended by a different DNA polymerase for short lengths (called Okazaki fragments). The RNA primers are eliminated and the DNA fragments are stitched together by a ligase enzyme (from Leman and Noguchi [4]).

thymine-bearing nucleotides. The nucleosomes in these stretches are enriched in the histone variant H2.AZ that attracts a methyl transferase enzyme responsible for methylating histone H4 at lysine 20 (H4K20me2). This modification is recognized and bound by the ORC [2]. The acetylation of other lysines (H4K5ac, H4K8ac and H4K12ac) occurs in mammals at replication origins [3].

Because the two complementary strands of a DNA molecule are oriented in opposite directions (Chapter 1) and because DNA polymerase can only add nucleotides at the 3′ end of a growing chain, DNA replication is continuous along one of the two DNA strands and discontinuous along the other strand (Fig. 3).

The replication process terminates when converging replication forks from adjacent origins meet and the newly formed DNA molecules (consisting of one old strand and its copy) from one replisome are fused with the DNA molecules of the adjacent replisome (see Box 1).

Box 1

Termination of DNA replication: This is achieved when the newly replicated DNA from the leading strand of a fork that was progressing, say towards the left, encounters the last Okazaki fragment of the adjacent fork that is moving to the right. The RNA primer of this fragment is removed, the gap is filled by the DNA polymerase that is copying the leading strand and the two newly synthesized strands are ligated (Fig. 4). When this is achieved, the replisome is disassembled.

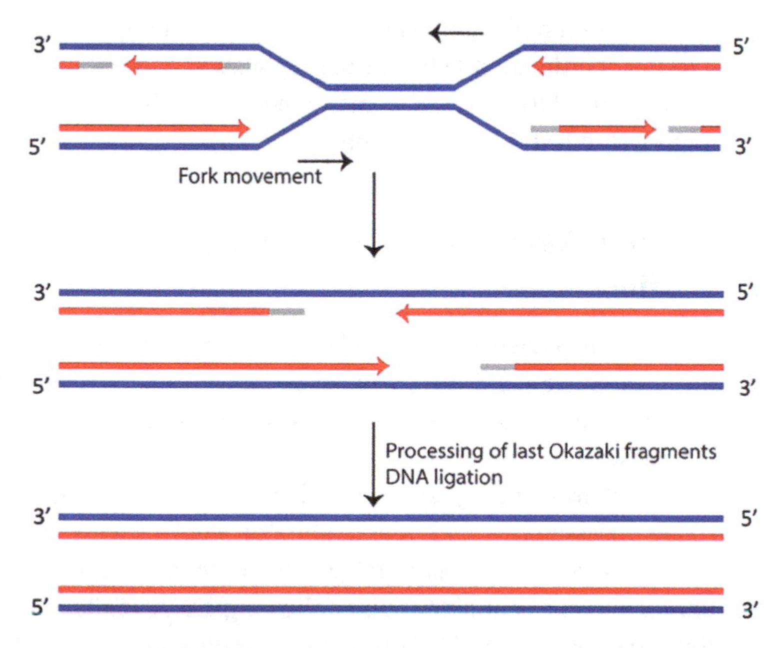

Fig. 4. Convergence of two adjacent replication forks. Okazaki fragments are indicated as a short gray segment attached to a short red arrow that represents a newly replicated segment of DNA. The leading strand replicated DNA of one replication fork meets the last Okazaki fragment of the lagging strand of the adjacent fork and, following removal of the last RNA primer, the gap is filled by extension of the leading strand replicated DNA and ligation of the two newly synthesized DNA fragments (modified from Moreno and Gambus, 2020[a]).

Note that the leading strand (5′ to 3′) in the replication fork that is moving towards the left becomes a lagging strand in the replication fork that is

(*Continued*)

Box 1 (*Continued*)

moving to the right because the replication has to proceed 3′ to 5′ and DNA polymerase can only add nucleotides at the 3′ end of the DNA copy that it is synthesizing.

---

[a]Moreno, S.P., and A. Gambus, Mechanisms of eukaryotic replisome disassembly. *Biochemical Society Transactions*, 2020. **48**(3): 823–836.

The timing of DNA replication is not uniform throughout the genome. In addition to the coordination between the replication and the transcription of regions that need to be active as mentioned above, heterochromatic regions are replicated last during the S phase.

B. Nucleosome Assembly Following Replication

Following replication, nucleosomes must be reconstituted along the newly formed molecules. Since there is now twice as much DNA present in the cell, the number of nucleosomes that were disassembled to allow replication to occur must be supplemented by a comparable number of nucleosomes constituted from newly synthesized histones. As mentioned in Chapter 4, the disassembly and reassembly of nucleosomes are the responsibility of chaperones; most notables among those involved in the reconstitution of the chromatin structure following DNA replication are *anti-silencing function 1* (ASF1), *chromatin assembly factor 1* (CAF1) and *nucleosome assembly protein 1* (NAP1), as well as the *facilitate chromatin transcription* (FACT) remodeling complex. A very important aspect of the reconstitution of chromatin is the re-establishment of epigenetic marks on the newly synthesized DNA molecules and on the newly assembled nucleosomal histones so that the pattern of gene function characteristic of the cell will be reproduced in the two daughter cells following mitosis [5]. The transmission of epigenetic marks through the cell cycle will be discussed in Chapter 9.

C. Errors in Replication and the Damage Response

Although DNA replication is a very precise process, errors can occur. Occasionally, the DNA polymerase inserts a wrong nucleotide or too many nucleotides, or fails to add nucleotides to the strand that it is synthesizing. Sometimes, ribonucleotides — the building blocks of RNA — are incorporated by mistake. In some cases, the DNA polymerases can recognize the misincorporated nucleotides, pause and remove them. In many cases, though, the errors just mentioned may cause the replication fork to stall and result in the occurrence of single-stranded DNA breaks (SSBs) or double-stranded breaks (DSBs) which can lead to mitotic arrest and cell death unless they are repaired. Several repair pathways exist to counteract the occurrence of DNA breaks. These pathways will be discussed in Chapter 8.

References

[1] Fragkos, M., *et al.*, DNA replication origin activation in space and time. *Nature Reviews Molecular Cell Biology*, 2015. **16**(6): 360–374.

[2] Long, H., *et al.*, H2A.Z facilitates licensing and activation of early replication origins. *Nature*, 2020. **577**(7791): 576–581.

[3] Che, Y.-J.C., *et al.*, Now open: Evolving insights in the role of lysine acetylation in chromatin organization and function. *Molecular Cell*, 2021. **82**(4): 716–727.

[4] Lehman, A.R., and E. Noguchi, The replication fork: Understanding the eukaryotic replication machinery and the challenges to genome duplication. *Genes*, 2013. **4**(1): 1–32.

[5] MacAlpine, D.M., and G. Almouzni, Chromatin and DNA replication. *Cold Spring Harbor Perspectives in Biology*, 2013. **5**(8): 1–22.

CHAPTER 8

How Does DNA Repair Occur in Chromatin?

In addition to the errors that occur during DNA replication, the genetic material is constantly subjected to a variety of external and internal agents that can lead to mutations, single- or double-strand DNA breaks or chromosomal rearrangements. The major categories of these agents are ultraviolet or ionizing radiation, chemical carcinogens and oxidative radicals.[1] Cells have at their disposal a number of DNA repair pathways that can be triggered to respond to specific categories of damage. These pathways usually involve DNA polymerases that are different from the polymerases involved in standard DNA replication [1].

A. Repair of Mispaired or Damaged Bases

Mispaired bases are recognized by the **mismatch repair** (MMR), **base excision repair** (BER) or **nucleotide excision repair** (NER) pathways. The proteins of the MMR system recognize the replication error and specifically cleave the new DNA strand to allow an enzyme (*endonuclease*) to remove the region of the error; a special DNA polymerase then fills the

[1]Oxidative radicals are oxygen-containing molecules with an uneven number of electrons that allows them to easily react with various organic substrates, such as lipids, proteins and DNA. They are the normal products of cellular respiration but, if their levels become excessive, they can cause damage to cellular components.

gap using the old DNA strand as a template and a *ligase* enzyme seals the strand break. BER is a repair pathway that removes damaged or altered DNA bases. The damaged nucleotide is removed and the correct nucleotide is added by another special DNA polymerase followed by ligation. There are two types of NER: one pathway acts on damaged DNA wherever it may occur in the whole genome and the other acts during transcription when the RNA polymerase stalls because of a lesion in the DNA template. The NER pathways deal with base gaps or bulky additions caused by chemical carcinogens or UV irradiation. These defects are detected by different sensing protein complexes that recruit the transcription factor TFIIH; in a manner similar to its formation of the transcription bubble, this factor uses its helicase and the energy liberated by breaking down ATP to unwind the DNA segment that contains the problem. The other components of the NER pathway function in a manner very similar to those involved in BER (Fig. 1).

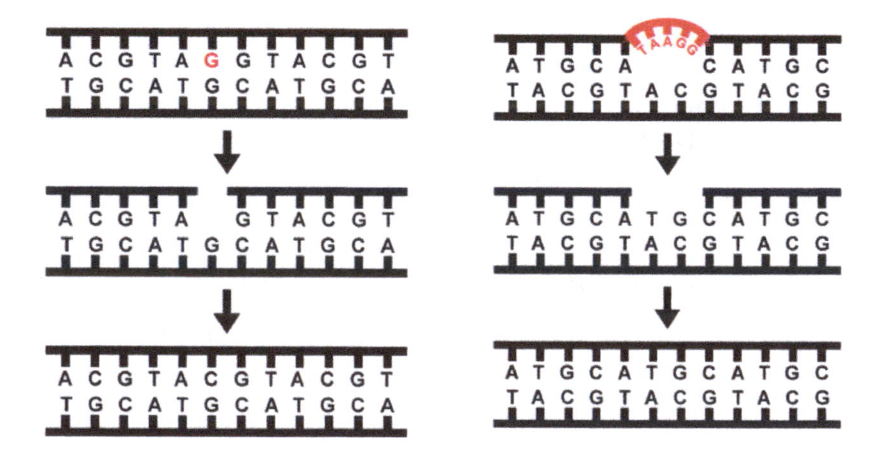

Fig. 1. Diagrams of the mismatch repair (MMR) and nucleotide excision (NER) DNA repair pathways. MMR (left diagram) corrects mispaired or damaged bases. Bulky DNA lesions or additions of extra bases are removed by the NER pathway (right diagram) that is initiated by unwinding the affected DNA region by the TFIIH transcription complex. The following steps, similar in the two pathways, involve excision by endonucleases and the addition of either a single nucleotide (BER) or the synthesis of a DNA segment followed by ligation.

B. Repair of DNA Breaks

Single-stranded breaks: Single-strand breaks are usually caused by errors in DNA replication, by oxidative damage or by ionizing radiation. If they are not repaired, they can become double-strand breaks and cause DNA replication or transcription to stop, or induce factors that cause cell death. The breaks are detected by special factors that recess the broken ends and leave a single-stranded gap. The gap is filled by the enzyme complexes and factors of the BER pathway.

Double-strand breaks: Double-strand breaks can be caused by a variety of external agents that include ionizing radiation, industrial chemicals or chemicals that are used for chemotherapy in cancer treatments, and internal agents such as reactive oxygen species.[2] They can also be caused by DNA replication through a single-strand break. They are repaired by two different pathways: the *homologous recombination* (HR) and the *non-homologous end-joining* (NHEJ) pathways. Which pathway will be used depends on the particular complex that recognizes the presence of the double-stranded break. These complexes recruit either the HR or the NHEJ repair components. The HR pathway uses some of the factors and enzymes that are responsible for chromosomal crossing over — the exchange of material that occurs between homologous chromosomes during meiosis (Fig. 2). The NHEJ pathway is initiated by a protein heterodimer (Ku) that recognizes the break and associates with both broken ends. Ku recruits the other components of the repair pathway that resect the ends, synthesize DNA to fill the gap and ligate the ends to complete the repair. In contrast to the HR pathway, NHEJ repair can often lead to mutations in the form of short nucleotide deficiencies.

Double-strand breaks are also transiently induced by enzymes (type II topoisomerases) that maintain the topological structure of DNA molecules by preventing them from getting underwound, overwound or tangled (see Box 1). Because they generate double-strand breaks, type II topoisomerases are inherently double-edged swords that have the ability to fragment the

---

[2] Reactive oxygen species are chemicals that contain super reactive forms of oxygen. They are the normal products of cellular respiration but, if their levels become excessive, they can cause *damage to DNA and other cellular components*.

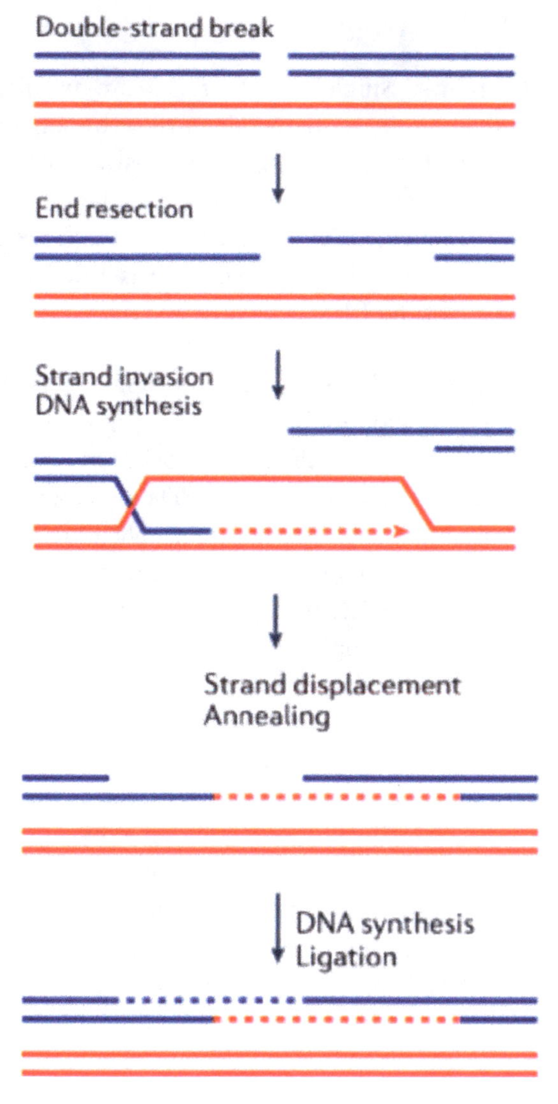

Fig. 2. Diagram of the homologous recombination (HR) repair pathway. Following recognition of the presence of a double-strand break, repair is initiated by resection of the broken ends which is necessary to produce 3′ ends that can be added to by DNA polymerase. Strand invasion allows the polymerase to have a complementary sequence to copy during DNA synthesis (modified from Sung and Klein [2]).

Box 1

The function of type II topoisomerases: To allow either transcription or replication to occur, the two strands of DNA molecules must be unwound. The formation and movement of a transcription bubble cause overtwisting of DNA downstream and undertwisting upstream; the movement of replication forks causes overtwisting ahead and intertwining of the two daughter molecules (precatenanes) behind the moving replisome (Fig. 3).

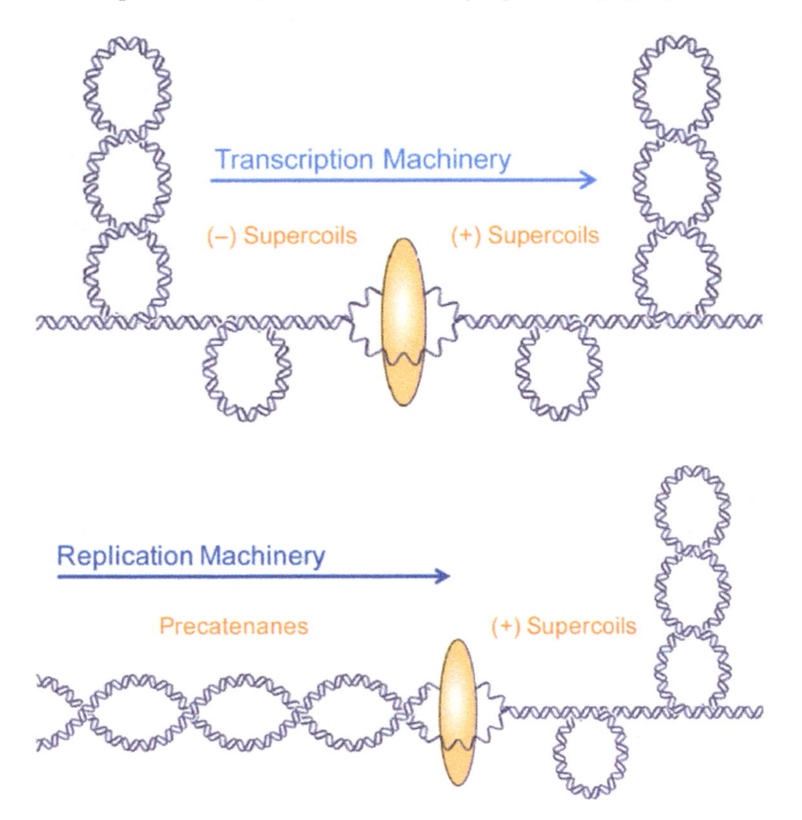

Fig. 3. Changes that can occur in the configuration of DNA molecules as a result of transcription or replication. The ovals represent either a transcription elongation complex (TEC) or a pair of replisomes (modified from Ashley *et al.*, 2017[a]).

(*Continued*)

Box 1 (*Continued*)

The unwinding process is the responsibility of a special type of enzyme, topoisomerase II.

This enzyme cuts both strands of the DNA at the same time (i.e., induces a double-strand break) and allows one double helical section of DNA to pass "through" another before re-ligating the two cut strands back together (Fig. 4).

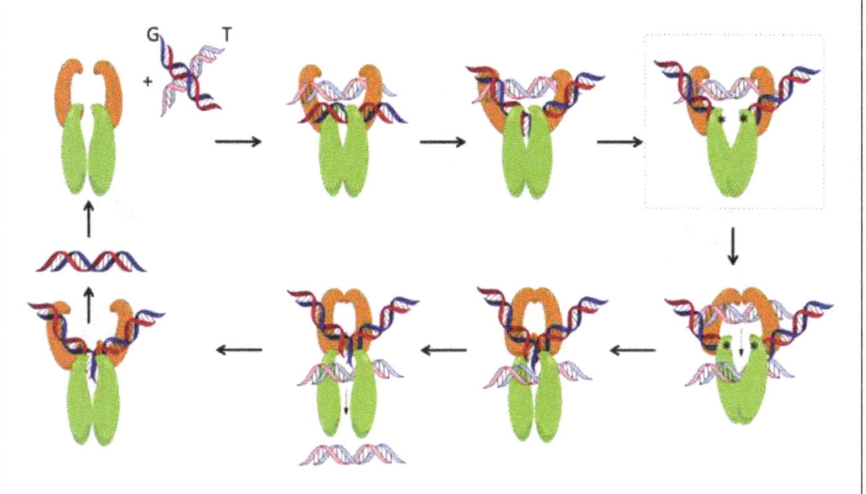

Fig. 4. Mechanism of DNA unwinding by type II topoisomerase. The enzyme binds to the DNA and causes a cut through both strands; another region of the enzyme captures the DNA at a more distant position from the cut and pushes it through the cut. Eventually, the two ends of the cut DNA are ligated back together (modified from Vann, 2021[b]).

---

[a]Ashley, R.E., *et al.*, Recognition of DNA supercoil geometry by mycobacterium tuberculosis gyrase. *Biochemistry*, 2017. **56**(40): 5440–5448.

[b]Vann, K.R., A.A. Oviatt, and N. Osheroff, Topoisomerase II poisons: Converting essential enzymes into molecular scissors. *Biochemistry*, 2021. **60**(21): 1630–1641.

genome. Such problems would arise if their ability to cleave DNA were not balanced by the re-annealing function.

C. Epigenetic Modifications Involved in DNA Repair

The various DNA damage repair pathways involve a variety of post-translational modifications of histones and repair proteins that enable the detection of the damage and recruit or activate the repair proteins. For example, histone acetylation and H3K79 methylation are thought to promote the NER pathway, and one of the first modifications associated with this pathway is the methylation of lysine 20 on histone H4 (H4K20me2). NER also relies on monoubiquitination of histone H2A. Both modifications attract the remodeling complexes that recognize the presence of the DNA lesion and repair it. Recently, the methylation of H3K36 appears to be associated with the NER pathway that is involved in repairing stalled RNA polymerases. The polyubiquitination of some of the repair complex components targets them for degradation once their function is completed [3–5].

Some of the best-studied epigenetic modifications involve the repair of double-strand breaks. A number of general chromatin modifications are associated with the occurrence of these breaks: if the breaks occur in euchromatic or actively transcribed regions, the chromatin is modified into a repressive state by the demethylation of H3K4me3 and the trimethylation of H3K9 and H3K27; if the breaks occur in heterochromatic regions, the chromatin is opened by the eviction of histone H1 and by histone acetylation to allow access to the break by the repair complexes ([6] and see Chapter 6). One of the first specific chromatin modifications found to be associated with double-strand DNA damage repair is the phosphorylation of a histone H2A variant. This variant, termed H2AX, is deposited throughout the genome by the FACT complex and, is phosphorylated along a region that spans the site of the double-strand break as soon as the latter is detected. H2AX replaces the canonical H2A histone in up to 25% of an organism's nucleosomes. Phosphorylated H2AX is monoubiquitinated and serves as a docking platform for members of the repair complex [7]. Another major determinant in DNA double-strand break repair is the methylation of lysine 20 on histone H4 (H4K20me2). Together with ubiquitinated H2AX, this modification provides a bivalent signal that allows the cell to decide whether to use the HR or NHEJ repair pathways [8].

References

[1] Chatterjee, N., and G.C. Walker, Mechanisms of DNA damage, repair and mutagenesis. *Environmental Molecular Mutagenesis*, 2017. **58**(5): 253–263.

[2] Sung, P., and H. Klein, Mechanism of homologous recombination: Mediators and helicases take on regulatory functions. *Nature Reviews Molecular Cell Biology*, 2006. **7**(10): 739–750.

[3] Borsos, B.N., H. Majoros, and T. Pankotai, Emerging roles of post-translational modifications in nucleotide excision repair. *Cells*, 2020. **9**(6): 1–15.

[4] Chitale, S., and H. Richly, DICER- and MMSET-catalyzed H4K20me2 recruits the nucleotide excision repair factor XPA to DNA damage sites. *Journal of Cell Biology*, 2018. **217**(2): 527–540.

[5] Selvam, K., *et al.*, Set2 histone methyltransferase regulates transcription coupled-nucleotide excision repair in yeast. *PLOS Genetics*, 2022. **18**(3): 1–23.

[6] Caron, P., E. Pobega, and S.E. Polo, DNA double-strand break repair: All roads lead to heterochROMAtin marks. *Frontiers in Genetics*, 2021. Article ID 34539757, 11 pages.

[7] Kelliher, J., G. Ghosal, and J.W.C. Leung, New answers to the old RIDDLE: RNF 168 and the DNA damage response pathway. *The FEBS Journal*, 2021. Article ID PMC8486888, 14 pages.

[8] Gong, F., and K.M. Miller, Histone methylation and the DNA damage response. *Mutation Research*, 2019. **780**: 37–47.

CHAPTER 9

Chromatin Fate in Normal Development

A. The Fate of Chromatin Modifications Through the Cell Cycle

During development, the diversification of cells into specialized tissues is initiated in the early embryo by the presence of *morphogens* (transcription factors, non-coding RNAs, etc.) differentially deposited into the developing egg often in the form of gradients. The presence of these morphogens and their different concentration in different regions of the early embryonic cell mass leads to the activation of particular genes in these regions, which, in turn, activate new genes, leading to differentiated cell types (see Box 1). In order to give rise to specialized tissues, differentiated embryonic cells must transmit the characteristics that they have acquired to their daughter cells. As discussed in Chapter 7, DNA replication necessitates the eviction of existing nucleosomes and the re-establishment of the pre-mitotic chromatin state by the deposition of old and newly synthesized histones. Since gene activity is regulated through epigenetic signals present in chromatin, the faithful transmission of the genetic program responsible for a particular cellular type consists of transmitting the epigenetic regulatory information and, therefore, requires the reproduction of the epigenetic marks from cells to daughter cells. Nucleosomes are disassembled immediately ahead of the DNA replication fork and reassembled

Box 1

Morphogen gradients and their function: Most morphogens are proteins that bind to receptors embedded in the membrane of cells. This binding initiates a cascade of molecular interactions that terminate in the nucleus to activate the transcription of specific target genes. A seminal and well-described example of morphogen activity is found in the early development of *Drosophila* embryos (Fig. 1). In these, as well as in most multicellular organisms, development begins with the establishment of the body axes of the future embryo.

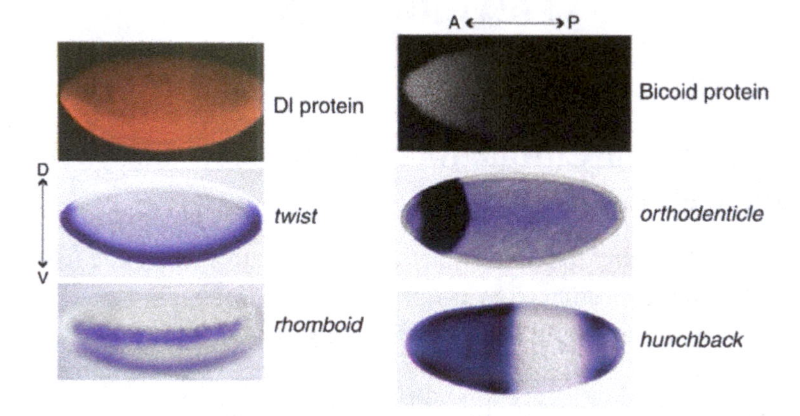

Fig. 1. Establishment of the dorso-ventral and antero-posterior body axes in a Drosophila embryo. The presence of a particular gene product is determined by using an antiserum that recognizes it and that is conjugated to an enzyme that catalyzes a color reaction (modified from Ashe and Briscoe, 2006[a]).

An activator molecule, deposited as a dorso-ventral gradient during the formation of the egg, results in the graded activation of the *Dorsal* (Dl) gene. A high concentration of the Dl gene product, in turn, activates the *twist* gene while a lower concentration activates the rhomboid gene. The differential presence of these gene products establishes what will be the dorsal and the ventral regions of the developing embryo. In a similar fashion, the *Bicoid* (Bcd) gene product is a transcription factor whose differential distribution induces the *orthodenticle* (ort) and the *hunchbac*k (hb) genes along the future embryonic antero-posterior axis.

[a]Ashe, H.L., and J. Briscoe, The interpretation of morphogen gradients. *Development*, 2006. **133**(3): 385–394.

behind the fork using both parental and newly synthesized histones. Parental histones retain the modifications (methylations, acetylations, etc.) that were present prior to DNA replication; as they remain in close proximity to their original DNA location, their incorporation into the newly formed nucleosomes provides the basis for re-establishing the epigenetic information of that particular genomic region in daughter cells. The restoration of the histone marks characteristic of repressed chromatin (H3K27me3 and H3K9me3) on newly synthesized histones is facilitated because the appropriate methyltransferases recognize these modifications on the parental histones and modify the neighboring, newly synthesized histones. Mention should be made that post-translational modifications that occur on particular core histones are conserved when the latter are replaced by histone variants, for example, when H3 is replaced by H3.3 during transcription. Following nucleosome disassembly during DNA replication, these histone variants are part of the pool of parental histones that are used to reconstitute the chromatin of daughter cells. The propagation of active chromatin histone marks is less understood; although it is possible that the incorporation of old histones with their transcription marks (H3K4me3, H3K36me3 and H3K79me3) on the newly synthesized DNA molecules promotes the resumption of gene activity in daughter cells [1], there are no demonstrated cases of the transmission of a specific activated state via a specific epigenetic modification. Recently, *prions* (see Section C) have been shown to transmit the active state of particular genes through cell divisions and across generations in yeast cells.

The other major epigenetic modification — DNA methylation — is a key player in conveying the parental cell's program of gene activity to daughter cells. Methylated or hydroxymethylated cytosines play a critical regulatory role in genome activity (discussed in Chapter 4). Therefore, it is necessary for the pattern of DNA methylation in cells to be transmitted accurately to daughter cells [2]. Following replication, each of the two new DNA molecules contains one old strand on which the methylated cytosines are present and a new strand with unmethylated cytosines. This "half-methylated" situation is recognized by a *maintenance methyltransferase* enzyme that restores the fully methylated status on the replicated DNA (Fig. 2).

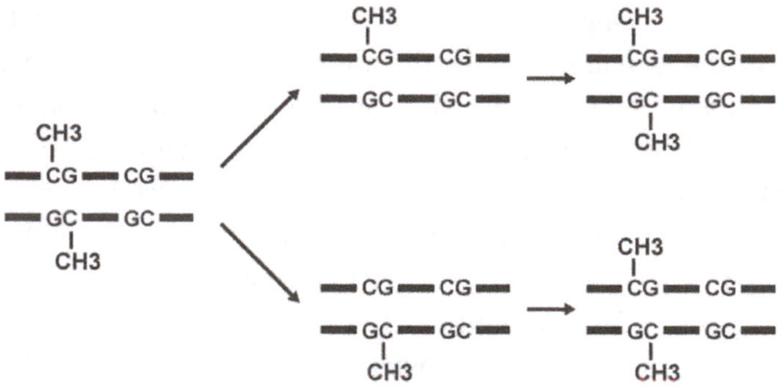

Fig. 2. Diagram illustrating the function of the maintenance DNA methyl transferase, DNMT1. Following DNA replication, the half-methylated site is recognized and is fully methylated.

B. Inheritance of Epigenetic Modifications

Following the seminal hypothesis of Conrad Waddington (discussed in Chapter 1), evidence has accumulated substantiating the effect of environmental stimuli, experienced during the formation of gametes by the parents and during early developmental stages, on the future behavior and health of an individual [3]. External factors include nutrition, some maternal diseases and exposure to toxins or to alcohol, nicotine and illegal drugs. These factors affect the normal epigenetic profile that occurs during the formation of eggs and sperm, and that directs the development of the fertilized egg into a multicellular, highly differentiated organism.

The germline — the source of future eggs or sperm — is derived from a stem cell population of *primordial germ cells* (PGCs) that are present during very early embryogenesis and that develop into male or female lineages depending on the sex of the gonad (ovary or testis). Mature sperm and eggs form following the onset of puberty. Specific epigenetic modifications occur during the formation of eggs and sperm [4]: in these gametes, most of the DNA methylation present is erased and DNA is newly methylated at imprinted loci and in regions rich with transposable elements (discussed in Chapter 4). In sperm, the majority (but, significantly, not all) of the histones are replaced by sperm-specific proteins termed *protamines.*[1] Small non-coding RNAs (miRNAs and piRNAs) are involved in the formation of both sperm and eggs; these RNAs are selectively expressed at different times during the differentiation of the PGCs. Exposure to toxic substances or to unfavorable dietary conditions can lead to changes in the pattern of DNA methylation and histone modifications during the formation of male or female gametes, in the pattern of histone replacement by protamines during sperm formation and in the composition of the miRNAs. The resulting epigenetic modifications alter gene expression during offspring development and often do not have consequences until adult life when they can lead to the onset of tumors, of metabolic defects such as obesity and diabetes, of cardiovascular diseases or of chronic respiratory and neurobehavioral defects [5]. This type of transmission of epigenetic information from parents to their offspring is referred to as *intergenerational epigenetic inheritance.*

An important question of current research is whether the epigenetic changes induced by exposure to some environmental condition, and that are responsible for the health outcome of offspring, are transmitted to subsequent generations that are not exposed to the causative condition. Such epigenetic changes are often called *epimutations*. The inheritance of some epigenetically based traits in model organisms can persist for only a few generations or can appear to be permanent. In humans, evidence for this type of inheritance, referred to as *transgenerational inheritance*, is still scarce for obvious reasons. In order to demonstrate the persistence of an epigenetic trait induced by exposure to some environmental factor, the trait must manifest itself in the great-grandchildren of an affected female or the grandchildren of an affected male. The reason is that the children are subjected to the factor during pregnancy (*in utero*) and so were their gametes, which form very early during embryonic development and give rise to the affected female's grandchildren. In the case of an exposed male, only the sperm that give rise to his children were affected by the environmental factor [6]. Therefore, information on the nutritional and clinical status of great-grandmothers and grandfathers of adult individuals would

---

[1] The remaining histones retain both active and repressive histone marks on enhancers and promoters that are involved in development. This observation supports the contention that epigenetic information can be passed from father to offspring.

be required. Furthermore, it is often difficult to distinguish between genetic (the actual effect of different gene alleles) and epigenetic contributions to a particular phenotype. Nevertheless, the substantial amount of data derived from a variety of experimental animals evokes the possibility of transgenerational inheritance in humans.

Among the first established cases of transgenerational inheritance were the effects of toxic compounds that interfere with the functions of hormones: *endocrine hormone disruptors*. Early studies that established the occurrence of this phenomenon involved the consequences of exposure to vinclozolin, a fungicide which affects estrogen function, and methoxychlor, a pesticide that was used after DDT was banned. Male rats exposed *in utero*, i.e., developing in gravid females that were exposed to these compounds, exhibited abnormal spermatogenesis; this defect was transmitted until the fourth generation together with an increased occurrence of liver cancer, kidney disease and immunological problems [7,8].

A relatively common neurobehavioral disorder in the human population is *attention deficit hyperactivity disorder* (ADHD). Using mice that had been induced with ADHD by prenatal exposure to nicotine, the hyperactive behavior was transmitted to the third generation of offspring even though only the founder generation had been exposed to nicotine [9].

The role of epigenetic transmission has been investigated in numerous farm animals [10]. While most of these studies have focused on the transmission of effects from treated parents to their offspring (intergenerational inheritance), some have studied subsequent generations. A large study on the effect of diet on the epigenetic transgenerational inheritance of phenotypic characters has been carried out in pigs. Boars were fed a diet that contained high levels of compounds involved in the synthesis of S-adenosylmethionine, the principal substrate for DNA methylation. The F2 generation (grandchildren) of treated boars had significantly less fat than controls. This phenotype was correlated to an increased level of DNA methylation of dozens of genes active in muscle tissue, the liver and the kidneys [11].

The transmission of phenotypic characteristics that occur in later generations is the result of epigenetic modifications that are induced as a result of exposure to environmental factors during crucial windows in germ cell development of the founder generation. The role of DNA methylation or of histone modifications has been implicated in various animal experiments. Epimutations that consist of changes in DNA methylation face the same problem that concerns imprinted alleles: in order to be transmitted to subsequent generations, they must either avoid the waves of demethylation that occur during development (see Chapter 6) or be precisely remethylated. Not surprisingly, transgenerational inheritance can be correlated to genetic mutations, if the latter alter the mechanisms responsible for establishing epigenetic modifications; the effect of some of these gene mutations was found to persist in subsequent generations that, by chance, did not inherit the mutant alleles. For example, as a consequence of exposing parents to heat shock, a novel protein correlated to a decrease in histone methylation was found in the descendants' eggs and sperm [12].

Other types of factors that mediate epigenetic regulation — the different types of non-coding RNAs (described in Chapter 4) — are involved in transgenerational inheritance. An early remarkable demonstration of this involvement was the demonstration that ncRNAs could be isolated from the sperm of stressed male mice and could cause similar behavioral and metabolic alterations if they were injected into fertilized wild-type oocytes [13]. In other experiments, pesticide-induced changes in ncRNAs were present in the sperm of the male offspring (F1), grandchildren (F2) and great-grandchildren (F3) of treated gestating female rats exposed to vinclozolin [14]. Not surprisingly, other epigenetic changes including differentially methylated regions of the genome and regions of differential histone retention were found to be colocalized [15].

C. Prions

Although all of the epigenetic changes discussed so far — DNA methylation, histone modifications and non-coding RNAs — have been associated with chromatin components, a different class of self-perpetuating and transmissible modifications is linked, directly, to proteins. Most of these proteins are *prions* (proteinaceous infectious particles), misfolded proteins that induce their normally folded isoforms to misfold.[2] This mechanism of self-propagation of particular proteins has been found in simple organisms such as yeast and has been studied by introducing it into invertebrate and vertebrate models (*Caenorhabditis*, *Drosophila*, zebrafish and mice). First discovered as the infective agent responsible for scrapie, a disease of sheep and goats, and kuru, a rare disease formerly found in the New Guinea population, prions are implicated in several neurodegenerative diseases in humans. The cause of these diseases is the tendency of misfolded proteins to aggregate and form fibrous deposits or plaques. In addition to these pathological effects, prions are emerging as ubiquitous proteins that control a variety of biological traits [16].

Prions can be transmitted through the cell cycle; this process involves specific chaperones that fragment the aggregates into "seeds" that are passed on to daughter cells and initiate a new round of protein isoforms misfolding. Although no cases have been described of their transgenerational inheritance in multicellular eukaryotes, a particular protein, transmitted to daughter cells, has been shown to induce an activated chromatin state in yeast cells [17].

References

[1] Stewart-Morgan, K.R., N. Petryk, and A. Growth, Chromatin replication and epigenetic cell memory. *Nature Cell Biology*, 2020. **22**(4): 361–371.

[2] Kim, M., and J. Costello, DNA methylation: An epigenetic mark of cellular memory. *Experimental and Molecular Medicine*, 2017. **49**(4): 1–8.

[3] Safi-Stibler, S., and A. Gabory, *Epigenetics and the developmental origins of health and disease: Parental environment signalling to the epigen*ome, critical time windows and sculpting the adult phenotype. *Seminars in Cell and Developmental Biology*, 2020. **97**: 172–180.

---

[2] Following their synthesis, all proteins fold into one or more specific spatial conformations that are driven and maintained by a number of non-covalent interactions between particular amino acids. The proteins that give rise to prions contain a disorganized region that can assume a new conformation responsible for their infective nature. This misfolding can be caused by mutations that alter the amino acid sequence or by environmental factors that alter the physiology of cells. Prions are characterized by their resistance to *proteases*, enzymes that break down proteins.

[4] Ben Maamar, M., E.E. Nilsson, and M.K. Skinner, Epigenetic transgenerational inheritance, gametogenesis and germline development. *Biology of Reproduction*, 2021. **105**(3): 570–592.

[5] Fleming, T.P., *et al.*, Environmental exposures around conception: Developmental pathways leading to lifetime disease risk. *International Journal of Environmental Research and Public Health*, 2021. **18**(17): 1–18.

[6] Skinner, M.K., Environmental stress and epigenetic transgenerational inheritance. *BioMed Central Medicine*, 2014, Article ID: PMC4244059, 5 pages.

[7] Anway, M.D., *et al.*, Epigenetic transgenerational actions of endocrine disruptors and male fertility. *Science*, 2005. **308**(5727): 1466–1469.

[8] Anway, M.D., *et al.*, Endocrine disruptor vinclozolin induced epigenetic transgenerational adult-onset disease. *Endocrinology*, 2006. **147**(12): 5515–5523.

[9] Zhu, J., *et al.*, Transgenerational transmission of hyperactivity in a mouse model of ADHD. *The Journal of Neuroscience*, 2014. **34**(8): 2768–2773.

[10] Thompson, R.P., E. Nelson, and M.K. Skinner, Environmental epigenetics and epigenetic inheritance in domestic farm animals. *Animal Reproduction Science*, 2020. **220**(106310): 1–10.

[11] Braunschweig, M., V. Jagannathan, A. Gutzwiller, and G. Bee, Investigations on transgenerational epigenetic response down the male line in F2 pigs. *PLoS ONE*, 2012. **7**(2), e30583: 1–9.

[12] Klosin, A., *et al.*, Transgenerational transmission of environmental information in *C. elegans*. *Science*, 2017. **356**(6335): 320–323.

[13] Gapp, K., *et al.*, Implication of sperm RNAs in transgenerational inheritance of the effects of early trauma in mice. *Nature Neuroscience*, 2014. **17**(5): 667–669.

[14] Ben Maamar, M., *et al.*, Alterations in sperm DNA methylation, non-coding RNA expression, and histone retention mediate vinclozolin-induced epigenetic transgenerational inheritance of disease. *Environmental Epigenetics*, 2018. **4**(2): 1–19.

[15] Beck, D., M. Ben Maamar, and M.K. Skinner, Integration of sperm ncRNA-directed DNA methylation and DNA methylation-directed histone retention in epigenetic transgenerational inheritance. *Epigenetics & Chromatin*, 2021. **14**(6): 1–14.

[16] Itakura, A.K., *et al.*, Widespread prion-based control of growth and differentiation strategies in *Saccharomyces cerevisiae*. *Molecular Cell*, 2020. **77**(2): 266–278.

[17] Harvey, Z.H., *et al.*, A prion epigenetic switch establishes an active chromatin state. *Cell*, 2020. **180**(5): 928–940.

CHAPTER 10

Stem Cells

A. Types and Characteristics

Different types of stem cells: The development of most multicellular organisms begins with a single cell — the fertilized egg — which, through successive mitotic divisions, produces an aggregate of genetically identical cells. Each of the first few embryonic cells that are produced is able to give rise to a whole individual and is said to be *totipotent*. As they continue to divide, embryonic cells become limited to producing particular cell types that constitute the *germ layers* of the developing embryo[1]; these cells, referred to as *stem cells*, are now considered to be *pluripotent*. Within each germ layer, groups of cells develop that can differentiate into the different types of cells that are present in a particular tissue; these cells are referred to as *multipotent* (Fig. 1).

Adult tissues that undergo continuous cellular turnover contain undifferentiated, multipotent stem cells. In vertebrates, examples of these tissues are the skin, the lining of the digestive track, the reproductive organs that produce eggs and sperm and the bone marrow[2] that is responsible for the production of all of the different types of blood cells (Fig. 2).

---

[1]Germ layers are layers of cells that form during embryonic development and that give rise to specific groups of different tissues in the growing organism. Multicellular organisms, from simple forms to vertebrates and including plants, have two or three germ layers (in vertebrates, the germ layers are the ectoderm, mesoderm and endoderm).

[2]The existence of stem cells was discovered by three Canadian investigators who demonstrated that some cells derived from mouse bone marrow could differentiate into a variety of different cell types [2].

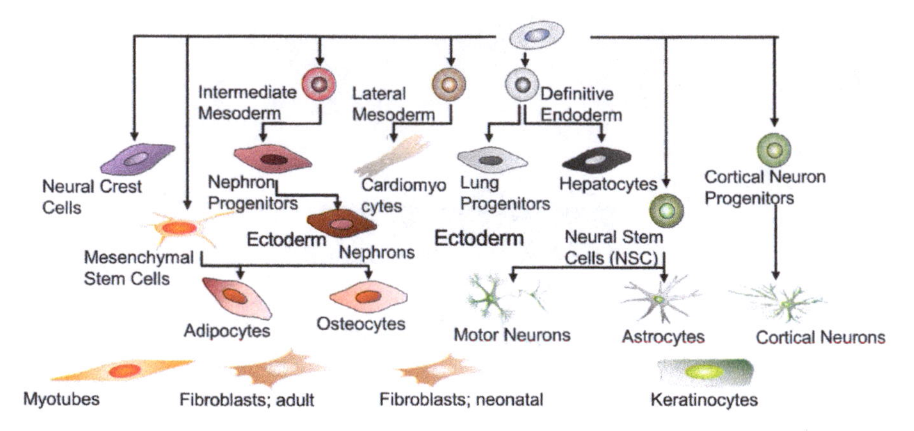

Fig. 1. In the early human embryo, totipotent cells give rise to the three germ layers (mesoderm, endoderm and ectoderm) that in turn produce the progenitor cells of all of the tissues found at birth (modified from Grosch *et al.* [1]).

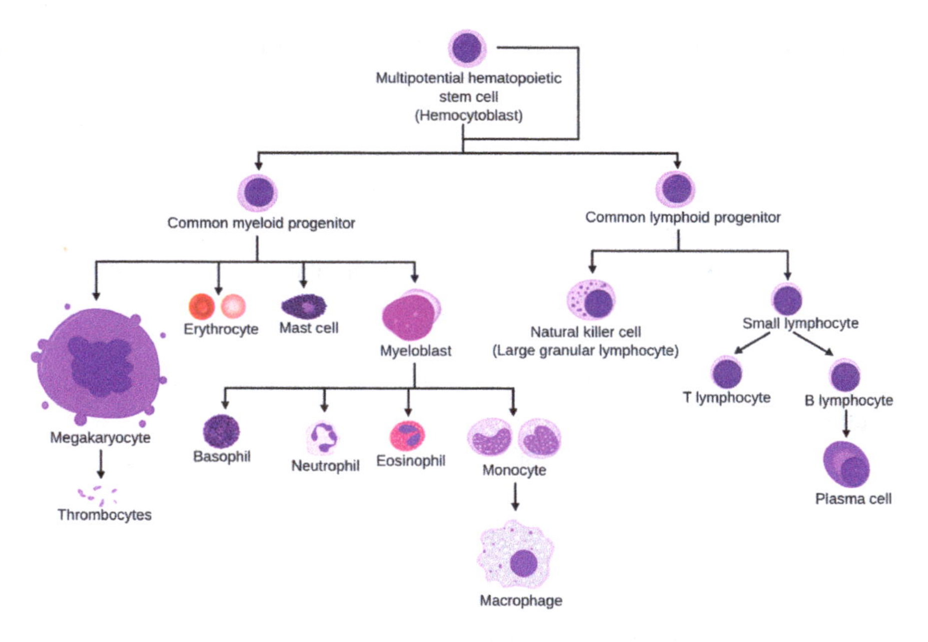

Fig. 2. Formation of blood cells from multipotent stem cells in the bone marrow. The stem cells divide to reproduce themselves or to give rise to cells that have reduced differentiation capability (progenitor cells). In turn, progenitor cells can give rise to a single type or several types of differentiated blood cells (by A. Rad and M. Häggström. CC-BY-SA 3.0 license).

In general, the maintenance of the pluripotent state of stem cells is the responsibility of several specific transcription factors among which Oct4, Sox2, NANOG and KLF4 are the most important, as well as some gene products that are overexpressed in tumors, such as c-Myc[3] (see the following).

The niche concept: The differentiation of adult multipotent stem cells depends on their location within the tissue. Within this location, termed the *stem cell niche*, stem cells are able to self-renew by mitotic division but are also directed to differentiate according to the tissue's need for cell replacement. The niche controls the stem cells' fate by direct cell–cell interactions through adhesive molecular signals and through diffusible factors [3].

Chromatin characteristics: Stem cells ensure that their two defining characteristics — self-renewal and the ability to differentiate into different cell types — are maintained by a set of coordinated gene activations modulated by changes in chromatin structure [4]. Except for the regions of the genome that consist of constitutive heterochromatin, the chromatin of stem cells is generally more open than in differentiated cells. This allows stem cells to be poised to respond to differentiation signals. Embryonic stem cells contain specific remodeling complexes, such as particular members of the SWI/SNF family, that are necessary for altering the status of chromatin during differentiation. Many stem cell gene promoters are marked by the presence of H3K56ac, a histone mark with which some of the stem cell-specific transcription factors (Oct4, Sox2 and NANOG) colocalize. Many stem cell promoters are *bivalent*, i.e., they are marked with both active (H3K4me3) and repressive (H3K27me3) histone modifications; removal of the repressive mark would allow the rapid expression of genes involved in differentiation. The H3K27me3 repressive mark is established by the action of the PcG1 repressive complex (see

---

[3]As is the case for so many factors and other entities involved in the molecular biology of chromatin, these transcription factors have rather bizarre names: Oct4 (octamer-binding transcription factor 4), Sox2 (SRY-box transcription factor 2), KLF4 (Krüppel-like factor 4), NANOG (this name is derived from a word in Irish mythology) and c-Myc (Avian myelocytomatosis viral oncogene homolog).

Chapter 6); during the activation of bivalent promoters, PcG proteins are evicted by SWI/SNF complexes and H3K27me3 is cleared through the action of a particular chaperone (ASF1A, *Anti-silencing function 1A*). Following the onset of differentiation, the genes that must remain silent in a particular cell type are rendered inactive by the action of CHD4, a chromatin remodeler that removes the acetyl group from H3K27ac and thereby allows its methylation to H3K27me3 by the PRC2 inactivating complex, a characteristic of inactivated chromatin.

Another feature of the chromatin of pluripotent stem cell is the clustering of enhancers, termed *super-enhancers*, that facilitate the differentiation into particular cell types by simultaneously activating a particular group of genes. In undifferentiated stem cells, these upper-enhancers are occupied by the pluripotency regulators Oct4, Sox2 and NANOG.

B. Nuclear Reprogramming

Nuclear reprogramming consists of altering a genetic state characteristic of a particular cell type into that of a different cell type, usually less differentiated. There are two major purposes to this activity. The first is to clone animals by *stem cell nuclear transfer* (SCNT); the second is to generate experimentally *induced pluripotent stem cells* (iPSCs) to be used for medical regenerative purposes and plant propagation.

Organismal cloning was first accomplished in frogs. John Gurdon was able to transform the nucleus of a highly differentiated cell (from the gut of a frog's embryo or tadpole) into a totipotent nucleus by inserting it into a frog egg from which the nucleus had been removed. Such eggs were able to develop to the tadpole stage [5]. Since then, it has been successfully applied to a variety of animal species including a number of mammals. Coupled with the ability to alter genomes by the CRISPR method (discussed in Chapter 2), some of the major goals have been to generate farm animals with favorable characteristics, to rescue endangered species and to reproduce animal models of human disease for medical research purposes. The first and most famous animal to be produced by nuclear transfer was a sheep named Dolly [6]. Since then, over 20 mammalian species have been cloned including a monkey, the first non-human primate [7]. In spite of these successes, the cloning process has been inefficient with only a very small percentage of eggs with transplanted nuclei developing to term. The causes of this inefficiency are exclusively epigenetic: the DNA methylation status and the array of histone modifications that are responsible for the genomic activity in the donor cell's nucleus have to be modified to resemble those present in a naturally fertilized egg; although factors present in the egg cytoplasm can achieve this process, it is often incomplete or inaccurate. It is not surprising that most of the research devoted to improving cloning by SCNT consists of improving epigenetic reconstruction [8].

The feasibility of plant cloning was established by the landmark observation that whole carrot plants could be regenerated from segments of vascular tissue. The process of producing whole plants from single somatic cells is termed *somatic embryogenesis*. Plant cells in culture can be induced to totipotency by exposing them to auxin, an essential "growth hormone" of plants, by inducing particular transcription factors (such as Lec1 or Lec2) or by interfering with the presence of the polycomb repressive complexes in order to recapitulate the generally more open chromatin of stem cells [9].

Given the great potential of stem cell therapy, the harvesting of *embryonic stem cells* (ESCs) has been the focus of a major research effort. ESCs are obtained from very early embryos (four to five days after fertilization, in the case of human ESCs) or from the tissues of slightly older embryos. The supply of ESCs is limited and, in the case of humans, is subject to ethical questions. For these reasons, mesenchymal stem cells (MSCs) and laboratory-induced pluripotent stem cells (iPSCs) represent promising alternatives. Initially isolated from bone marrow in vertebrates, MSCs can be obtained from umbilical cord blood, placenta and amniotic fluid [10].

Induced pluripotent stem cells (iPSCs): The production of pluripotent stem cells in the laboratory was made possible by the discovery that a very small set of transcription factors were sufficient to cause differentiated cells to revert to pluripotency. It was known for some time that the introduction of single transcription factors could transform one type of differentiated somatic cells into another type. Shinya Yamanaka tested the expression of some of these factors in different combinations; he discovered that the following transcription factors, Oct4, Sox2 and Klf4, together with c-Myc, a factor known to be upregulated in a number of tumors, could transform mouse embryonic or adult fibroblasts into pluripotent cells [11]. In order to demonstrate that the cells were now pluripotent, they were injected into mice where they formed teratomas (masses of differentiated tissues originating from all three embryonic germ layers). Another commonly used means of assessment is the injection of induced pluripotent cells into early mouse embryos to see if they lead to the birth of chimeric individuals, i.e., of mice that contain tissues derived from the injected cells [11]. Since Yamanaka's seminal discovery, a variety of iPSC lines have been generated and they have been induced to differentiate into almost all known vertebrate specialized cells and tissues [12].

A variety of alternative methods for generating iPSCs that are faster and more efficient have been and are continuously being sought. Some of these modify the epigenetic landscape of the cells that are being manipulated by targeting histone modifications or DNA methylation, or by supplementing the culture medium with certain chemicals, in order to improve the effectiveness of the Yamanaka transcription factors. Other methods rely on treatment of cells with microRNAs [13].

C. Relevance of Stem Cells to Human Health

Stem cell technology is one of the more exciting developments in current biomedical research with two major applications: (i) the establishment of cell cultures from the tissues of affected individuals for the molecular study of disease etiology and for drug discovery and (ii) regenerative medicine. Three sources of cells have been used: embryonic stem cells (ESCs), adult stem cells (ASCs) and iPSCs. ESCs have been used in clinical trials for the treatment of such conditions as heart disease, neurodegenerative diseases, spinal cord injuries and diabetes. The problems encountered in these trials are ethical (the use of human embryos) and immunological (rejection of the implanted cells). These problems are moot in the case of iPSCs which can be generated by culturing cells from adult donors and reverting them to pluripotency in the laboratory; when reintroduced into the same donor, the iPSCs are recognized as self and are immunologically neutral. Nevertheless, there are additional pitfalls that must be considered. If following *in vitro* differentiation of the iPSCs to the appropriate cell type some pluripotent cells remain, following transplantation these cells could give rise to tumors. Furthermore, during the process of culturing, spontaneous genetic errors may occur, such as chromosomal abnormalities or gene mutations. To circumvent this problem, whole-genome sequence analyses can be performed prior to the introduction of cells into the patient. While the direct injection method of treatment is usually performed for regenerative therapy, encapsulating the cells in implanting devices can be used for treatment of metabolic deficiencies. Such encapsulation can prevent the potential tumor formation by the undifferentiated cell contaminants. An example of these treatments, both of which have been used for the treatment of type I diabetes, involves the use of pancreatic cells derived from differentiated iPSC cells (Fig. 3).

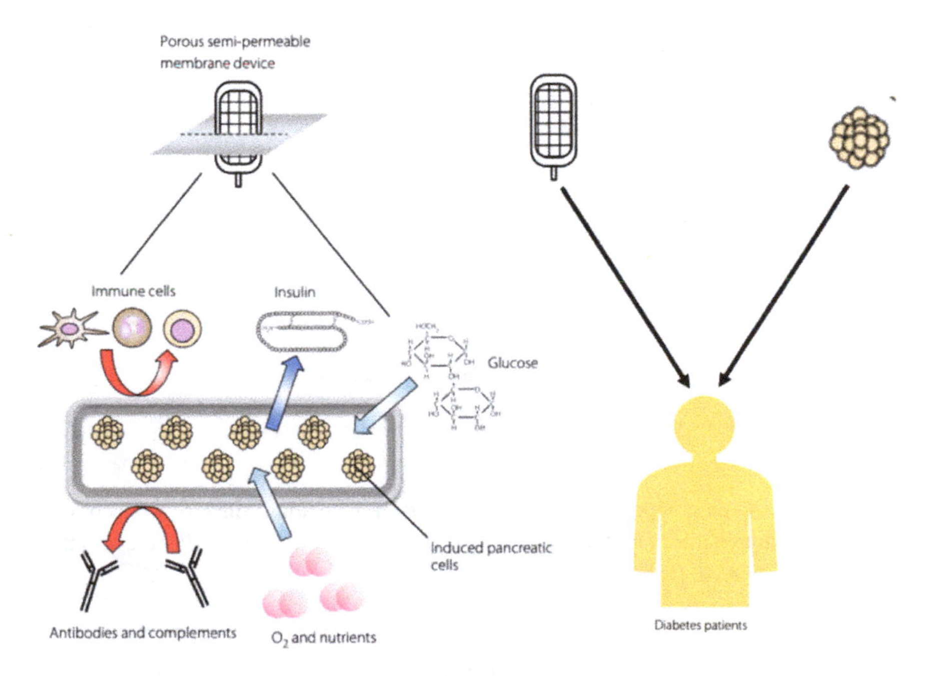

Fig. 3. Pancreatic cells, derived from iPSCs, differentiated *in vitro* to produce insulin can be encapsulated in a device with a semi-permeable membrane that allows the passage of nutrients necessary for the cells' maintenance and the passage of secreted insulin but prevents access to antibodies and killer immune cells. Stem cell-derived pancreatic cells can also be injected into the patient. In this particular case, they can be introduced into the liver by injecting them into the major liver blood vessel (modified from Kondo *et al.* [14]).

Another advantage of iPSCs is that they can be differentiated *in vitro* into diseased cells that replicate the cells of affected patients (Fig. 4). This experimental approach can create excellent cellular models for determining the molecular basis of the disease, for discovering new drugs and for testing therapies (see Box 1).

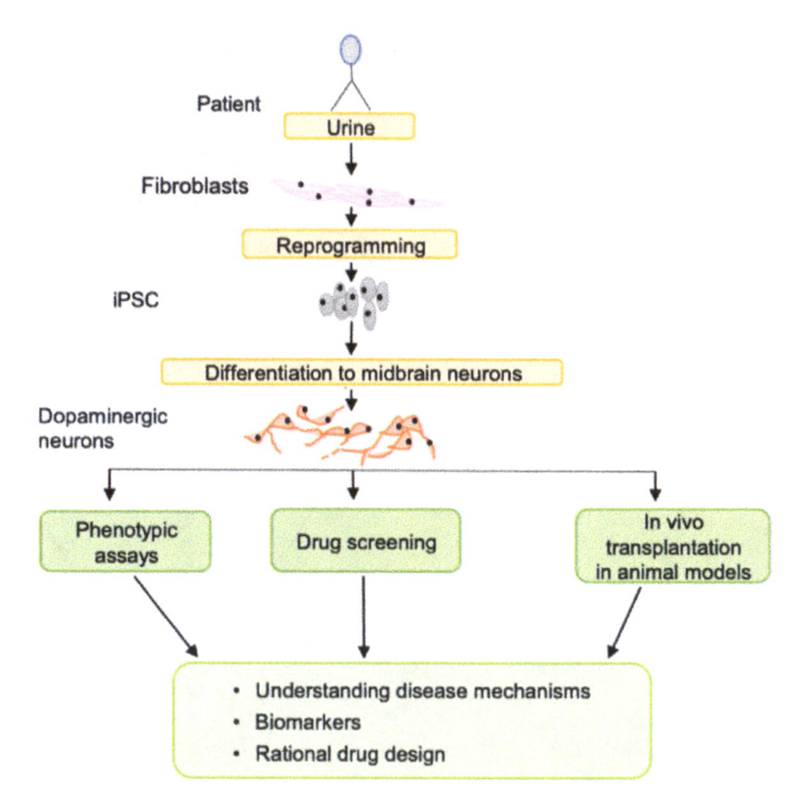

Fig. 4. Flow diagram representing the steps involved in creating cellular models of diseases. The example illustrates the use of iPSCs to study the pathology of Parkinson's disease (from Bose *et al.* [15]).

Box 1

Examples of iPSC-derived disease studies: Initially, iPSC-derived disease models were two dimensional, i.e., consisted of single-layered cell cultures. Recently, efforts have been devoted to develop differentiation protocols that

Box 1 (*Continued*)

result in the formation of three-dimensional masses of cells, so-called organoids, that better reflect the cellular circumstances of the diseased organ in the whole organism.

Among the countless examples of this use of pluripotent cells[a] is the study of cardiomyopathy, a disease of the heart muscle. The *in vitro* model consists of isolating iPSCs from patients and differentiating them into heart muscle cells (cardiomyocytes). The molecular genetic analysis of these cells revealed that a particular genetic mutation was responsible for the main symptoms including oversized and abnormal cell structure, poor contractions and abnormal calcium cycling. Drug testing revealed that compounds that blocked the cellular calcium channels were effective in alleviating the symptoms.

As mentioned in the text (see Fig. 4), iPSCs derived from cells of Parkinson's disease patients have been used to determine the effects of the different mutations responsible for the familial and the sporadic instances of the disease. A number of compounds were found to ameliorate the disease phenotype exhibited by the cell models, potentially leading to new therapeutic strategies.

Neurons derived from iPSCs produced from cells derived from patients with Alzheimer's disease develop the intracellular protein tangles (*amyloid* β peptides) and extracellular protein plaques (*tau* protein) characteristic of the disease. This model has been used to screen for different drugs that prevent the formation of these abnormal protein structures, among which are compounds that inhibit the phosphorylation of tau.

In addition to models of neurodegenerative diseases, iPSCs have been used to study neurodevelopmental problems whose etiology has been very difficult to unravel, such as autism spectrum disorder (ASD). Both genetic mutations and environmental factors are suspected to play a role in the onset of ASD. The genes that have demonstrated some involvement affect synaptic transmission, chromatin remodeling, protein metabolism and the cellular actin filaments' skeleton. Establishing iPSC models from patients with various mutations has enabled the parsing of genetic and environmental effects and the screening of different drugs to address them individually.

---

[a]Nicholson, M.W., *et al.*, Utility of iPSC-derived cells for disease modeling, drug development, and cell therapy. *Cells*, 2022. **11**(11): 1–17.

References

[1] Grosch, M., *et al.*, Nucleus size and DNA accessibility are linked to the regulation of paraspeckle formation in cellular differentiation. *BioMed Central Biology*, 2020. **18**(1): 1–19.

[2] Till, J.E., and E.A. McCulloch, A direct measurement of the radiation sensitivity of normal mouse bone marrow cells. *Radiation Research*, 1961. **14**: 213–222.

[3] Pennings, S., K.J. Liu, and H. Qian, The stem cell niche: Interactions between stem cells and their environment. *Stem Cells International*, 2018. Article ID: PMC6204189, 3 pages.

[4] Klein, D.C., and S.J. Hainer, Chromatin regulation and dynamics in stem cells. *Current Topics in Developmental Biology*, 2020. **138**: 1–71.

[5] Gurdon, J.B., The developmental capacity of nuclei taken from differentiating endoderm cells of *Xenopus laevis*. *Journal of Embryology and Experimental Morphology*, 1960. **8**: 505–526.

[6] Wilmut, I., *et al.*, Viable offspring derived from fetal and adult mammalian cells. *Nature*, 1997. **385**(6619): 810–813.

[7] Liu, Z., *et al.*, Cloning of macaque monkeys by somatic cell nuclear transfer. *Cell*, 2018. **172**(4): 881–887.

[8] Wang, X., *et al.*, Epigenetic reprogramming during somatic cell nuclear transfer: Recent progress and future directions. *Frontiers in Genetics*, 2020. **11**: 1–13.

[9] Su, Y.H., *et al.*, Plant totipotency: Insights into cellular reprogramming. *Journal of Integrative Plant Biology*, 2021. **63**(1): 228–240.

[10] Ntege, E.H., H. Sunami, and Y. Shimizu, Advances in regenerative therapy: A review of the literature and future directions. *Regenerative Therapy*, 2020. **14**: 136–153.

[11] Takahashi, K., and S. Yamanaka, Induction of pluripotent stem cells from mouse embryonic and adult fibroblast cultures by defined factors. *Cell*, 2006. **126**(4): 663–676.

[12] Karagiannis, P., *et al.*, Induced pluripotent stem cells and their use in human models of disease and development. *Physiological Reviews*, 2019. **99**(1): 79–114.

[13] Liu, G., *et al.*, Advances in pluripotent stem cells: History, mechanisms, technologies, and applications. *Stem Cells Reviews and Reports*, 2020. **16**(1): 3–32.

[14] Kondo, Y., *et al.*, iPSC technology-based regenerative therapy for diabetes. *Journal of Diabetes Investigation*, 2018. **9**(2): 234–243.

[15] Bose, A., G.A. Petsko, and L. Studer, Induced pluripotent stem cells: A tool for modeling Parkinson's disease. *Trends in Neuroscience*, 2022. doi: 10.1016/j.tins.2022.05.001. p. 1–13.

CHAPTER 11

The Consequences of Chromatin Dysregulation: Non-malignant Pathologies

As discussed in previous chapters, a major function of chromatin is to regulate the retrieval and function of the genetic information encoded in the DNA, in time and space, i.e., in different regions and cell types during development and adult life. The epigenetic modifications responsible for this function of chromatin can be altered in response to environmental influences and organisms are able to tolerate a certain level of variation in epigenetic regulation, just as they can tolerate the presence of some allele variations in the genetic blueprint. Nevertheless, it is clear that major changes in genetic regulation can have pathological consequences. These changes can be the result of alterations in epigenetic regulation caused by environmental factors or by genetic mutations in the genes that encode the enzymes and the non-coding RNAs responsible for this regulation [1]. A caveat in evaluating the association between epigenetic modifications and disease is the need to determine if the modifications cause or contribute to the symptoms or if they are a consequence of the disease process. In either case, epigenetic changes are an integral part of the etiology of the disease syndrome and can be targeted for therapeutic intervention.

A. Neurodegenerative Diseases

In humans, a particularly well-defined epigenetics-based neuropathy is the *Rubinstein–Taybi syndrome* (RSTS), a relatively rare condition characterized by growth retardation, intellectual and behavioral impairment and a variety of organ system malformations [2]. It is caused by a dominant mutation in one of two genes. The mutations can occur in the germline (the group of cells that form eggs and sperm) and be inherited, or they can occur within the cells of a developing individual. In the former case, RSTS is said to be *familial*, and in the latter case, *sporadic*. The two genes whose mutations can cause the syndrome are closely related and encode two transcription coactivators with intrinsic histone acetylating activity that play a crucial role in transcription initiation of a very large number of genes.[1] For these reasons, the effects of the mutations are not limited to the nervous system.

Another neurodevelopmental disorder directly associated with abnormal epigenetic regulation is the *Rett syndrome* (RTT). Affected individuals present with a complex set of abnormalities that can include motor dysfunction, loss of speech and respiratory difficulty, smaller head development (microcephaly) and epileptic seizures [3]. The primary cause of RTT is a mutation in the gene encoding a protein that normally binds to methylated CpG sites in the genome.[2] Because the gene is on the X chromosome, the mutation is lethal to males during fetal development; females with two X chromosomes have one normal allele of the gene and, therefore, are able to survive and exhibit the disease. Modern genome sequencing techniques have identified a number of other genes whose mutations cause RTT-like neurodevelopmental disorders. In most cases, these genes [e.g., FOXG1 (Forkhead box G1) and CDKL5 (cyclin-dependent kinase-like 5)] are involved in the same molecular pathways as the methylated DNA-binding protein.

---

[1]The two gene products involved in the transcriptional coactivation of many transcription factors are CREB (a protein that binds to the C-AMP-response element, CRE) and EP300 (this protein also binds to CREB). Both proteins are histone acetyltransferases that acetylate all four core histones.

[2]The methyl-CpG binding protein 2 (MECP2) specifically binds to the methylated CpG dinucleotides and recruits repressor proteins that inhibit gene transcription.

Alzheimer's disease (AD) is a slow-progressing, dementia-producing condition characterized by memory, cognitive and physical motion losses. The vast majority of hospitalized patients have a form of the disease referred to as late-onset. The causes are the interactions among genetic, epigenetic and environmental factors. The heritability of late-onset AD is relatively high. The disease presents with the formation of extracellular protein plaques and intracellular protein tangles.[3] Mutations in the genes that encode these proteins or in the genes responsible for the enzymes that transform them into pathological molecules have been identified. In addition, mutations in 20 additional genes increase the risk of developing the disease. Yet, it has been difficult to distinguish AD from other, similar neurodegenerative diseases. Recently, a number of epigenetic modifications have emerged as reliable biomarkers [4]. Although changes in chromatin modifications of specific genes were inconclusive, DNA methylation and hydroxymethylation have been shown to increase in particular brain regions of AD patients while decreasing in other regions and loss of heterochromatin leading to aberrant gene expression accompanies disease progression. With respect to histone modifications, a large number of chromatin sites are differentially marked with H3K27ac or with H3K9ac and there is a decrease in H4K16ac in different regions of AD brains in comparison to controls. Lastly, a number of specific miRNAs are dysregulated as a feature of the disease.

An interesting aspect of the etiology of AD is the involvement of modifications in the mitochondrial DNA of patients (see Box 1). Several mutations in the mitochondrial genome have been reported and a number of studies have implicated changes in DNA methylation and hydroxymethylation.

Parkinson's disease (PD) is another neurodegenerative disease that involves both genetic mutations and epigenetic changes [5]. PD affects the motor system leading to rigidity and paralysis, and causes cognitive and behavioral problems that manifest as dementia in advanced cases. It is characterized by the formation of protein plaques in selective neurons of

[3]The plaques consisting primarily of *amyloid* β peptides and the intra-neuronal tangles consisting of the protein *tau* are responsible for neural dysfunction and degeneration.

Box 1

Mitochondria biogenesis and heredity: Mitochondria are membrane-bound cellular organelles that are responsible for the synthesis of ATP by aerobic respiration (i.e., by the use of oxygen for oxidative phosphorylation). Found in the cytoplasm of most eukaryotic organisms, mitochondria contain their own genome consisting of a circular molecule of DNA, present in two to ten copies per mitochondrion, that encodes various enzymes for the respiratory cycle, 22 transfer RNA genes and a special form of ribosomal RNA. In addition, there are over 1,000 genes present in the nucleus that are involved in mitochondrial structure and function. In most multicellular organisms, the mitochondria of an individual are inherited from its mother through the egg cytoplasm. In most tissues, there are tens to hundreds of mitochondria per cell.

Although mitochondria have a version of most of the nuclear DNA repair pathways, mitochondrial DNA is mutated at a very high rate. When a mutation arises in a cell, the biogenesis of new mitochondria will give rise to a population of organelles, some of which may contain copies of the mutation in different proportions, due purely by chance (Fig. 1).

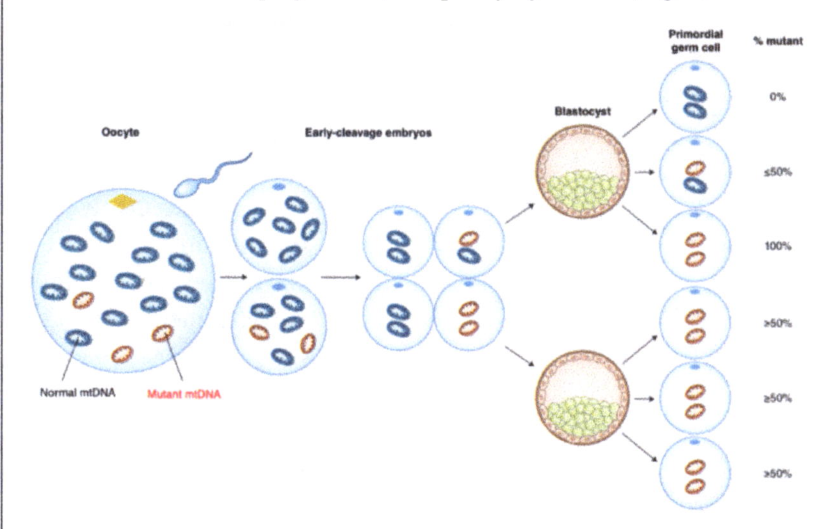

Fig. 1. Inheritance of mutated mitochondria. An oocyte contains a few copies of a mitochondrion with a mutation in its DNA. As the fertilized egg gives rise to early cleavage of female embryos (two-celled and then four-celled), the mutant-bearing mitochondria are distributed by chance. As development proceeds and the germ cells that will produce the next generation are formed, once again by chance, some oocytes will contain only the mutated mitochondria (modified from Wallace, 2017[a]).

Box 1 (*Continued*)

The fertilization of an oocyte that by chance contains only, or a high proportion of, mutated mitochondria would give rise to an individual whose energy metabolism is greatly impaired.

---

[a]Wallace, D.C., Mitochondrial genetic medicine. *Nature Genetics*, 2017. **50**(12): 1642–1649.

the brain resulting in the eventual loss of these neurons.[4] A number of mutations in different genes, including mutations in lysosomal proteins, have been associated with the onset or severity of PD (lysosomes are sub-cellular organelles that are responsible for the digestion of macromolecules, of old cell parts and of invading microorganisms).

An interesting epigenetic change is associated with PD. In normal individuals, the general level of DNA methylation has been shown to change proportionately as a function of age, with approximately half of the monitored sites becoming methylated and half losing their methylation; these changes are highly correlated with chronological age ([6] and further discussed in Chapter 12). In PD patients, the rate of methylation changes is significantly faster than in healthy individuals. This accelerated aging of the DNA methylome is also a characteristic of Alzheimer's disease and of Huntington's disease (discussed in the following). Global H3K27 regulation has been strongly implicated in PD with a decrease in methylation and an increase in acetylation of this lysine that affect the status of enhancers and deregulate the expression of the plaque-forming protein gene and other genes. *Several mi-RNAs that* have been shown to down-regulate the plaque protein are present in reduced levels in PD patients.

Huntington's disease (HD) inflicts movement, cognition and psychiatric disorders in patients; it is caused by a dominant mutation that leads to the overproduction of an oversized protein toxic to neurons in specific regions

---

[4]The plaques, referred to as Lewy bodies, are aggregates of the α-*synuclein* protein. These plaques form in specialized nerve cells that synthesize dopamine, a neurotransmitter molecule that affects the function of diverse regions of the brain.

of the brain[5] and to skeletal muscles and heart function problems. Epigenetic changes correlated with the disease are an increase in H3K9me3-silenced chromatin domains and a sequestering and redistribution of one of the major histone acetylating factors leading to changes in transcription [7]. In addition, several miRNAs, some of which are expressed specifically in neuronal cells, are dysregulated in HD [8].

B. Imprinting Diseases

As previously discussed (Chapter 4), genomic imprinting is an epigenetic mechanism that allows only one of the two alleles of a small number of genes to be expressed during development and, in many cases, in the adult organism. Imprinting is regulated by the differential DNA methylation of *imprint control elements* (ICEs) that, in turn, direct the methylation of those alleles that must not be expressed. ICEs regulate the imprinting of clusters of genes present on the same chromosome. All imprint marks must be erased early during the formation of gametes so that, during sperm production, a male can establish the imprint that will allow expression of only the maternal allele in the offspring, and vice versa.

Errors in the single-allele expression of imprinted genes can result from the deletion or duplication of ICEs or of the active parental alleles. They can also arise from errors during gamete formation that could lead to an offspring with both chromosomes of a particular pair coming from the same parent. The failure of homologous chromosomes to separate during the first meiotic division or of sister chromatids to separate during the second meiotic division (referred to as *non-disjunction,* see Chapter 2) can produce gametes that have two copies of a chromosome. If combined with a normal gamete from the other parent at fertilization, the resulting embryo would be *trisomic* for the chromosome, often a lethal condition.

---

[5]The mutation, located in the *Huntingtin* gene, causes the expansion of a trinucleotide (CAG) present in the gene's DNA sequence; CAG is the codon for the amino acid glutamine; this leads to the production of a longer protein which is toxic to certain neurons, if the expansion exceeds 40 copies (i.e., if a chain of 40 or more glutamines is added to the amino acid sequence of the protein). Acetylating enzymes are attracted to aggregates formed by the abnormal protein, affecting the levels of H3K27ac in different regions of the genome (7).

If, by chance, the chromosome from the other parent was lost during early development, the individual would wind up with both members of a chromosome pair from the same parent and be said to exhibit *uniparental disomy*. The misexpression of imprinted genes accounts for over a dozen imprinting disorders in humans, among which are the following examples.

Prader–Willi syndrome (PWS) and *Angelman syndrome* (AS) are characterized by early decrease in muscle tone and delay in development; PWS patients exhibit endocrinological and behavioral problems, while AS patients develop epilepsy in their early life. In PWS, approximately two-thirds of the cases involve the deletion of all or some of five genes and a cluster of small non-coding RNAs that are normally expressed from the paternal alleles; one-third of the cases are due to maternal uniparental disomy. In AS, a maternal deletion or a mutation of a single gene, normally expressed from the maternal allele, is present in over two-thirds of the patients; the remainder cases are caused by paternal uniparental disomy or mutations in the ICE [9].

Because of its dependence on DNA methylation, genomic imprinting is susceptible to environmental factors that alter this epigenetic mark during development (discussed in Chapter 9). While high levels of exposure to toxic chemicals[6] can have a drastic effect on imprinted genes and therefore lead to early lethality, low levels of exposure can cause adverse effects during development or increase the risk of adult disease [10].

C. Metabolic Syndrome Diseases

The *metabolic syndrome* consists of several biochemical problems including an increase in internal (visceral) fat deposits, high levels of blood glucose and high blood pressure; it can lead to the development of

[6] Some of the most widespread toxins are bisphenol A that is used in the production of plastic food and drink containers and can leach out and be ingested, phthalates that are used in the production of PVC products, toys, food packaging and personal care products, pesticides such as vinclozolin that is used to treat fruits and vegetables (it is currently banned in the U.S.), and dioxins that are produced as a result of burning fossil fuels but also forest fires and volcanic eruptions [10].

obesity, cardiovascular disease and diabetes. The DNA methylation status of several genes involved in lipid metabolism and a decrease in the level of particular histone-deacetylating enzymes have been shown to trigger the syndrome. Another important epigenetic factor whose deregulation can cause the onset of metabolic syndrome is miRNA. As previously discussed (Chapter 4), these small non-coding RNA molecules are involved in the transcription of a large number of genes and in the translation of protein-coding messenger RNAs, including as expected genes that encode DNA and histone modifying enzymes. In diabetes, the level of some of these mi-RNAs is affected in the cells responsible for insulin production [11].

One of the more important environmental factors associated with metabolic diseases is the diet and the epigenetic modifications that it can induce through its effect on the digestive track microbiome.[7] The digestive system of vertebrates and many other multicellular organisms is populated by a highly diverse number of bacterial, fungal and viral species. Microbial metabolites can affect all of the pathways of epigenetic regulation and, thereby, affect a number of metabolic diseases [12]. Much of the data supporting this contention are derived from animal studies. DNA and histone methylation are affected by the level of synthesis of the S-adenosylmethionine (the methyl group donor) from precursors, some of which are contributed or augmented by bacteria. Although the specific biochemical pathways are still poorly understood, the microbiome metabolic activity also affects the level of histone acetylation by inhibiting the activity of histone deacetylases. Finally, the profiles of lncRNAs and miRNAs in digestive system epithelial cells and in some cells of the immune system are influenced by the presence of particular bacterial species. Through these epigenetic modifications, the gut microbiota can regulate genes involved in fat synthesis and facilitate the onset of obesity, insulin resistance, i.e., diabetes, and cardiovascular disease [13].

---

[7]The population of microorganisms that colonize the intestinal tract is referred to as the intestinal flora or *microbiota*; the term "microbiome" refers to the total genomes of the microbiota and the functions that these genomes encode. Colonization begins after birth and the microbiota can be affected by different factors including diet, antibiotics and infections.

References

[1] Cavalli, G., and E. Heard, Advances in epigenetics link genetics to the environment and disease. *Nature*, 2019. **571**(7766): 489–499.

[2] Van Gils, *et al.*, Rubinstein-Taybi syndrome: A model of epigenetic disorder. *Genes (Basel)*, 2021. **12**(7): 1–22.

[3] Xiol, C., *et al.*, Technological improvements in the genetic diagnosis of Rett syndrome spectrum disorders. *International Journal of Molecular Sciences*, 2021. **22**(19): 1–19.

[4] Perkovic, M.N., *et al.*, Epigenetics of Alzheimer's disease. *Biomolecules*, 2021. **11**(2): 1–38.

[5] van Heesbeen, H.J., and M.P. Smidt, Entanglement of genetics and epigenetics in Parkinson's disease. *Frontiers in Neuroscience*, 2019. **13**: 1–15.

[6] Xiao, F.-H., H.-T. Wang, and Q.-P. Kong, Dynamic DNA methylation during aging: A "prophet" of age-related outcomes. *Frontiers in Genetics*, 2019. **10**: 1–8.

[7] Kim, C., *et al.*, Non-cell autonomous and epigenetic mechanisms of Huntington's disease. *International Journal of Molecular Sciences*, 2021. **22**(22): 1–25.

[8] Dong, X., and S. Cong, MicroRNAs in Huntington's disease: Diagnostic biomarkers or therapeutic agents? *Frontiers in Cellular Neuroscience*, 2021. **15**: 1–10.

[9] Wang, T.-S., *et al.*, Clinical characteristics and epilepsy in genomic imprinting disorders: Angelman syndrome and Prader-Willi syndrome. *Tzu Chi Medical Journal*, 2020. **32**(2): 137–144.

[10] Robles-Matos, N., *et al.*, Environmental exposure to endocrine disrupting chemicals influences genomic imprinting, growth, and metabolism. *Genes*, 2021. **12**(8): 1–23.

[11] Ramzan, F., M.H. Vickers, and R.F. Mithen, Epigenetics, microRNA and metabolic syndrome: A comprehensive review. *International Journal of Molecular Sciences*, 2021. **22**(9): 1–20.

[12] Qin, Y., and P.A. Wade, Crosstalk between the microbiome and epigenome: Messages from bugs. *Journal of Biochemistry*, 2018. **163**(2): 105–112.

[13] Sharma, M., *et al.*, The epigenetic connection between the gut microbiome in obesity and diabetes. *Frontiers in Genetics*, 2020. **10**: 1–15.

CHAPTER 12

The Consequences of Chromatin Dysregulation: Aging and Cancer

A. Aging

Aging is a complex biological process that involves the deregulation of epigenetic marks and directives resulting in a progressive reduction in stem cells and a progressive increase in genome instability manifested as an increase in DNA damage. As time passes, errors occur during DNA replication that are not caught and repaired by the various DNA repair mechanisms available to cells (see Chapter 8); these errors accumulate and are supplemented by errors that result from exposure to endogenous (e.g., reactive oxygen species that are the byproduct of mitochondrial function) or exogenous environmental factors, such as ionizing radiation or various chemicals. Other major characteristics of aging are the shortening of the tips of chromosomes, and *cellular senescence* whereby, after so many divisions, cells stop proliferating due to severe DNA damage and the depletion of replication factors. Directly associated with aging is a greater susceptibility to certain chronic diseases, such as the metabolic and neurodegenerative diseases (discussed in Chapter 11), and cancer [1].

DNA methylation: Aging involves changes in DNA methylation patterns. A number of CpG sites that gain or lose methylation over time can be selected to measure the rate of aging. These changes constitute a form of epigenetic *chronological clock.*[1] Like most if not all epigenetic marks, DNA methylation is influenced by environmental factors; changes induced by such factors transform the purely chronological clock into a biological clock [2]. Recall that in Chapter 11, the chronological clock of Parkinson's disease patients, for example, was found to be much shorter than that of healthy controls, highlighting the difference between chronological and biological clocks.

Histone modifications: Aging also involves histone modifications and a reduction in the synthesis of histones and in the function of remodeling complexes. These changes result in unregulated gene expression and in an age-related loss of heterochromatin [3]. Another consequence of loss of heterochromatin is the reactivation of previously silenced transposable elements (discussed in Chapter 4) whose multiplication and reinsertion in the genome can cause mutations. Finally, given their ubiquitous role in gene regulation, it is not surprising that many miRNAs are differentially expressed during aging [4].

All of the epigenetic modifications just discussed that increase as a function of age clearly have a profound influence on the transcriptional pattern of the cells that compose the different tissues of an organism. It is, therefore, not surprising that transcriptomic aging clocks can be established [5]. These clocks based on gene expression data (determine by mRNA sequencing) predict the biological age of individuals with unprecedented accuracy.

Telomeres: An important aspect of the aging process is the shortening of *telomeres*. In all eukaryotes, each round of DNA replication leaves a

---

[1] Several age-related changes in CpG methylation were found to be conserved in a number of mammalian species.

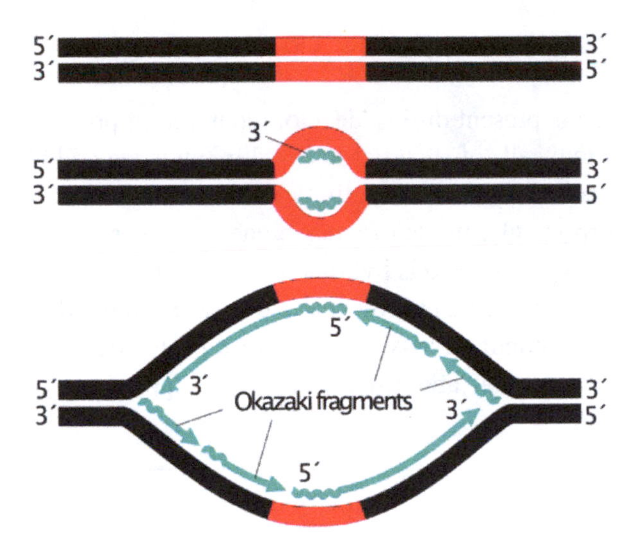

Fig. 1. Diagram of DNA replication. The red segment indicates the position of the origin of replication (see Chapter 7). As the replication bubble opens, the two replication forks that initiated at the origin move in opposite directions. The wavy green lines are RNA primers that have to be laid down because DNA polymerase can add nucleotides only at the 3′ end of an RNA or DNA molecule. Replication from the origin to the 3′ end of a DNA strand (upper right and lower left of the bottom figure) must occur in the form of a series of Okazaki fragments that occur in succession as the fork proceeds to the end of the strand. Once the RNA primer of the last Okazaki fragment is removed, it cannot be replaced by DNA, leaving a short single-stranded region.

single-stranded segment at the 3′ end of each chromosome (Fig. 1). Telomeres prevent the progressive degradation of these ends by enzymes that digest single-stranded DNA; they also ensure that these single-stranded ends are not mistaken for single-strand breaks by DNA repair pathways (discussed in Chapter 8). Telomeres achieve these goals by providing a cap for the ends of chromosomes consisting of a series of repetitive short nucleotide sequences to which special proteins are bound. The repetitive segment is laid down by an enzyme (*telomerase*) that uses an RNA sequence as a template for the DNA repeats that it synthesizes (see Box 1).

Telomerase is present during development and in postnatal and adult proliferating stem cells of such tissues as skin, intestine and blood, and in cells of the immune system; its activity is low or repressed in differentiated cells. Surprisingly, the telomerase gene promoter is highly methylated when it is expressed and is hypomethylated in those tissues where the enzyme is absent. As aging progresses, telomerase activity decreases in all tissues with a concomitant loss of telomere repeats, leading to stem cell depletion, chromosomal abnormalities and a suppression of a type of

Box 1

Telomere biogenesis: Telomerase is known as a "reverse transcriptase", i.e., an enzyme that uses RNA as a template for DNA synthesis; it is part of a complex that includes the template RNA and several accessory proteins. The RNA molecule pairs with a complementary sequence present at the single-stranded 3′ end of the newly replicated DNA molecule. This sequence is present because it was laid down during the synthesis of the telomere that followed the previous replication. It is single stranded because the RNA primer of the last Okazaki fragment, once digested, could not be filled in by DNA polymerase. The polymerase adds a few complementary bases to the 3′ end and then translocates and repeats the process (Fig. 2).

Following the release of the telomerase, the elongated strand is used as a template by a DNA polymerase to fill in the gap in the newly replicated strand. Of course, the polymerase has to use RNA primers that, when removed, will leave a single-stranded fragment at the telomerase elongated 3′ end. This single-stranded 3′ fragment loops back, invades the two-stranded region and pairs with a homologous sequence on the opposite strand forming a so-called T-loop (Fig. 3).

A series of different proteins bind to the DNA of the T-loop to form the *shelterin* complex that represses DNA repair mechanisms and generally protects the end of the chromosome. How the telomere is unfolded to allow the next round of DNA replication is not known at this time. One possibility is that it is resolved by the DNA helicase associated with the replisomes.

Box 1 (*Continued*)

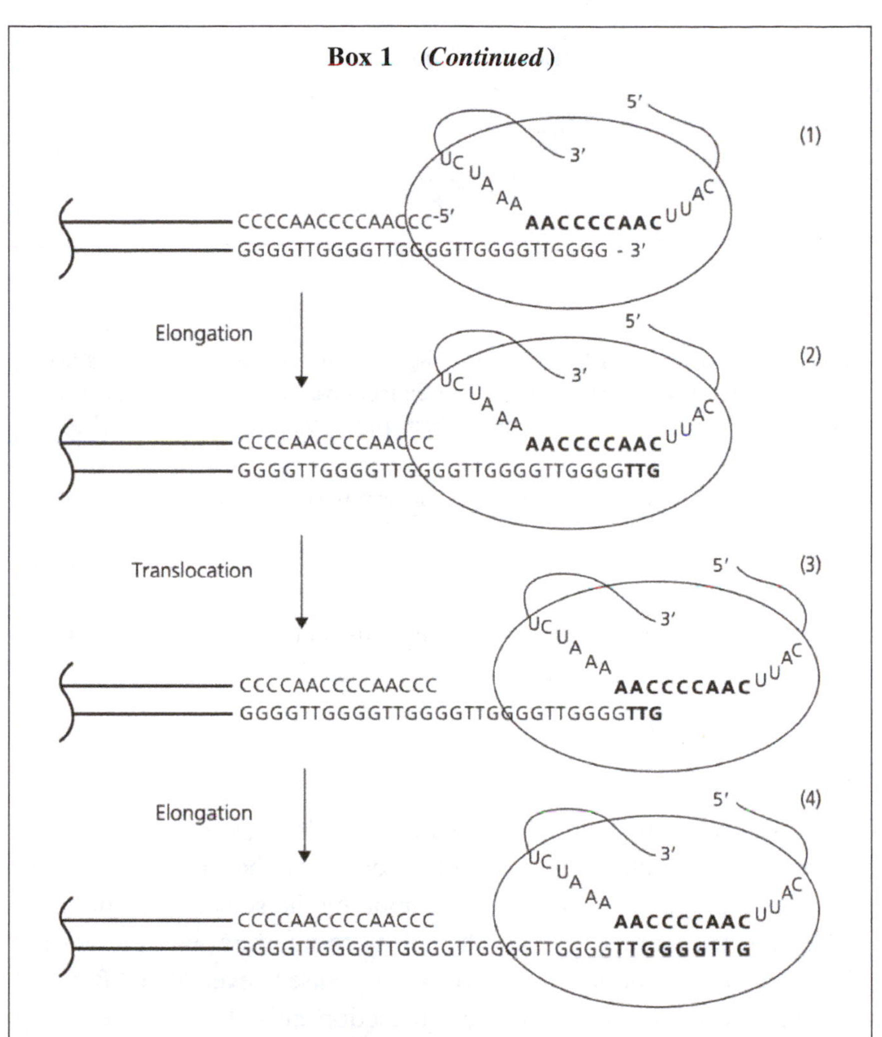

Fig. 2. Diagram illustrating the synthesis of a telomere. In this example, the sequence that is repeated many times to form the telomere is TTGGGG. (1) The RNA of the telomerase complex binds to its complementary sequence in the 3′ single-stranded end of the newly synthesized DNA strand. (2) The reverse transcriptase adds nucleotides to that strand. (3) The telomerase complex translocates, i.e., allows its RNA molecule to pair with the DNA sequence it just deposited. (4) Step 2 is repeated (from Greider and Blackburn, 1989[a]).

[a]Greider, C.W., and E.H. Blackburn, A telomeric sequence in the DNA of Tetrahyma telomeres required for telomere repeat synthesis. *Nature*, 1989. **337**(6205): 331–337.

(*Continued*)

Box 1 (*Continued*)

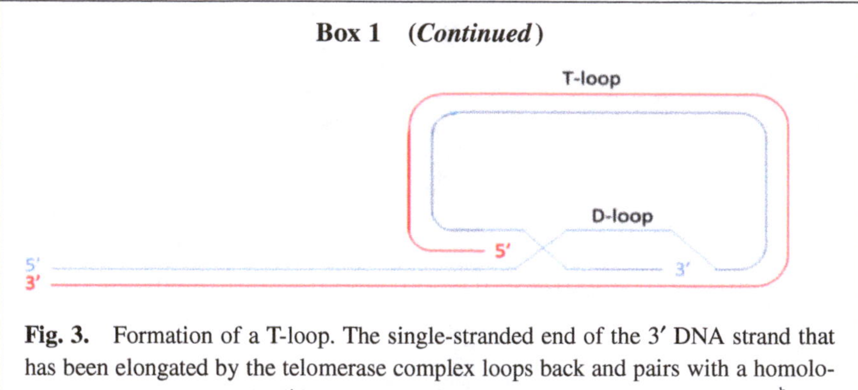

Fig. 3. Formation of a T-loop. The single-stranded end of the 3′ DNA strand that has been elongated by the telomerase complex loops back and pairs with a homologous sequence on the old 5′ DNA strand (from Turner, Vasu and Griffin, 2019[b]).

[b]Turner, K.J., V. Vasu, and D.K. Griffin, Telomere biology and human phenotype. *Cells*, 2019. **8**(1): 1–19.

histone deacetylases (*sirtuins*) that regulate certain proteins normally involved in delaying cellular aging [6].

B. Cancer

Cancer is the abnormal growth of cells within a tissue; these cells have the potential to invade other regions of the body. A number of hallmarks that characterize cancers have been agreed upon by the scientific community [7]. Among the more salient are the ability to interfere with the normal differentiation of multipotent stem cells or to cause a reversal in differentiated cells, the ability to avoid immune detection and cell death, the ability to replicate indefinitely, the ability to acquire a blood supply and the ability to spread from primary to other sites (metastasize). In addition, cancer cells exhibit DNA damage that causes them to undergo changes in their metabolism and allows them to maintain continuous cell growth. These characteristics are the result of mutations in *oncogenes* that, in their wild-type form (referred to as *proto-oncogenes*), promote cell growth and division, and *tumor suppressor genes* that normally limit cell growth and proliferation. Mutations that cause oncogenes to overexpress or mutations that inactivate tumor suppressor genes are detected in a large number of different types of cancers; as expected, these mutations alter the epigenetic programs characteristic of the normal cells. Consistent with this notion is the observation of a general genome-wide decrease in the level of DNA methylation, although in many cancers there is an increase in methylation of CpG islands associated with gene promoters of tumor suppressor genes and key developmental regulators.

Alterations in the cell cycle: The different steps or phases of the cell cycle are initiated and controlled by a specific set of factors[2] whose role is also to ensure that a given phase is properly completed before proceeding to the next phase; these factors act as checkpoints during the interphase period (when cell growth and DNA replication occur) and during the various stages of oncogenes and tumor suppressor genes' activity; the latter include the genes responsible for the various DNA repair pathways, since mutations in these genes would allow the accumulation of genomic errors and damage that are a characteristic of cancer cells.

Cancer stem cells: An important aspect of oncogenesis is the presence of cancer stem cells (CSCs) in almost all cancer types. As mentioned previously (Chapter 10), most tissues in multicellular organisms contain stem cells that are capable of self-renewal, i.e., of dividing to generate other stem cells, or that can differentiate into progenitor cells, which then give rise to sets of specialized cells characteristic of the tissue. There is substantial evidence that cancers originate from stem cells or from early progenitor cells that have acquired a number of particular mutations [8]. Such mutations — causing the overexpression of oncogenes or the inactivation of tumor suppressors — re-establish the ability to self-renew and prevent senescence and cell death. In these cells, the

[2]The different steps of the cell cycle are initiated and controlled by cyclins and cyclin-dependent kinases (CDKs). Each cyclin molecule partners with a particular CDK to form an active regulatory dimer. There are several checkpoints in the cell cycle but the main ones are the G1 checkpoint that ensures that the cell is ready to start DNA replication (S phase), the G2/M checkpoint that verifies that the cell is ready to enter into mitosis and the SAC checkpoint that monitors the proper formation of the spindle.

epigenetic programs responsible for self-renewal and differentiation have been altered.[3]

Cellular senescence: One aspect of regulation of the cellular life cycle of potentially oncogenic cells is *cellular senescence*, an arrest in proliferation caused by DNA damage usually induced by oxidative stress, ultraviolet or ionizing radiation. This damage triggers the induction of specific cell cycle control factors that induce irreversible cellular arrest and prevents the proliferation of these damaged cells that could cause tumor formation. It also induces the degradation of a histone H3 methyltransferase enzyme, responsible for the formation of H3K9me, a major histone isoform characteristic of inactive chromatin. The resulting effect is more open chromatin that allows the activation of genes responsible for the *senescence-associated secretory phenotype* (SASP) and the production of many secreted proteins including cell cycle control and growth factors. Mention should be made that the SASP plays a significant number of additional roles: it can induce the senescence of surrounding cells, and it can provide a tumor-inducing environment. In contrast, it can recruit cells of the immune system to the site of wounds, and it actually can serve as a tool for the formation of organs during embryonic development [9].

A major modification that allows many different types of cancer cells to avoid replicative senescence and to replicate indefinitely is to prevent telomere shortening by maintaining an active telomerase. This can be achieved through promoter mutations, promoter methylation or the action of super-activated oncogenes that encode transcription factors targeted to the telomerase gene promoter.

The structural organization of chromatin is affected in cancer cells: A number of heterochromatic regions that were normally associated with the nuclear periphery leave their location and coalesce to form new heterochromatin foci. This rearrangement brings genes that are near the heterochromatic regions in proximity, leading to the activation of genes that affect cell adhesion and of other cancer-related genes [10]. In addition, a significant proportion of all cancers (more than 20%) exhibit mutations

[3] From a clinical point of view, the greater the proportion of undifferentiated, primitive cells in cancer, the poorer the prognosis.

that inactivate or increase the function (so-called gain-of-function mutations) of chromatin remodelers. Since remodelers are associated with particular gene promoters and enhancers, loss-of-function mutations are likely to prevent the expression of tumor suppressor genes, while gain-of-function mutations can increase the accessibility of various genes including oncogenes [11].

Environmental influence: Although less extensively documented, the possibility that environmental influences may induce cancers by modifying epigenetic marks in the absence of genetic mutations is suggested by some existing cases. An example is a type of children's cancer that is induced by hypoxia (lack of adequate oxygen supply) and appears to lack any underlying genetic mutations [12]. The affected brain tissue exhibits a high level of replacement of methylated H3K27 by its acetylated form due to an increase in specific histone demethylating enzymes and an increase in acetyl-CoA, the substrate of lysine acetyltransferases. Hypoxia also induces transcription factors that bind the telomerase gene promoter and induce telomerase transcription [13].

References

[1] Saul, D., and R.L. Kosinsky, Epigenetics of aging and aging-associated diseases. *International Journal of Molecular Sciences*, 2021. **22**(1): 1–25.

[2] Li, A., Z. Koch, and T. Ideker, Epigenetic aging: Biological age prediction and informing a mechanistic theory of aging. *Journal of Internal Medicine*, 2022. doi: 10.1111/joim.13533: 1–12.

[3] Lee, J.-H., *et al.*, Heterochromatin: An epigenetic point of view in aging. *Experimental and Molecular Medicine*, 2020. **52**: 1466–1474.

[4] Kinser, H.E., and Z. Pincus, MicroRNAs as the modulators of longevity and the aging process. *Human Genetics*, 2020. **139**(3): 291–308.

[5] Meyer, D.H., and B. Schumacher, BiT age: A transcriptome-based aging clock near the theoretical limit of accuracy. *Aging Cell*, 2020. doi: 10.1111/acel.13320: 1–17.

[6] Chakravarti, D., K.A. LaBella, and R.A. DePinho, Telomeres: History, health, and hallmarks of aging. *Cell*, 2021. **184**(2): 306–322.

[7] Hanahan, D., Hallmarks of cancer: New dimensions. *Cancer Discovery*, 2022. **12**(1): 31–46.

[8] Clarke, M.F., Clinical and therapeutic implications of cancer stem cells. *New England Journal of Medicine*, 2019. **381**(10): 2237–2245.

[9] Ohtani, N., The roles and mechanisms of senescence-associated secretory phenotypes (SASP): Can it be controlled by senolysis? *Inflammation and Regeneration*, 2022. **42**(1): 1–8.

[10] Sati, S., *et al.*, 4D genome rewiring during oncogene-induced and replicative senescence. *Molecular Cell*, 2020. **78**(3): 522–538.

[11] Clapier, C.R., Sophisticated conversation between chromatin and chromatin remodelers, and dissonances in cancer. *International Journal of Molecular Sciences*, 2021. **22**(11): 1–37.

[12] Michealraj, K.A., *et al.*, Metabolic regulation of the epigenome drives lethal infantile ependymoma. *Cell*, 2020. **181**(6): 1329–1345.

[13] Dogan, F., and N.R. Forsyth, Telomerase regulation: A role for epigenetics. *Cancers*, 2021. **13**(6): 1–25.

CHAPTER 13

Research Prospects

A mantra espoused by most scientific researchers states that it is usually impossible to predict how basic research can be translated into practical applications relevant to human health. There are, of course, numerous examples of particular biological phenomena, initially investigated to satisfy the desire to better understand them, that have led to medical interventions and treatments. The fundamental, basic knowledge of genetic inheritance and epigenetic regulation, acquired over time by studying not only model experimental organisms but an endless variety of other life forms, discussed in the first 10 chapters of this primer, has led to the relevant observations on the human conditions discussed in Chapters 11 and 12. It is reasonable to wish and expect that current and future basic research will continue to extend our understanding of human development and health. A few among the many areas under investigation are presented in the following.

Stem cells: One of the promising subjects is the use of stem cells to correct genetic and degenerative disorders [1]. Stem cells (discussed in Chapter 10) can be simply isolated from a particular tissue and used within the same tissue, or they can be created in the laboratory (iPSCs) and primed for differentiation into particular cell types. This therapeutic approach requires very precise knowledge of stem cell maintenance and differentiation, as well as the development of efficient protocols for their clinical delivery. It also requires overcoming a number of challenges that include tumorigenicity, immunogenicity and heterogeneity [2]. One of the advantages of growing pluripotent stem cells is their ability to grow and proliferate in culture; but this characteristic must be very carefully controlled because if the cells keep proliferating after transplantation, the result may be tumor formation. Tumors may also result if the factors that induced the cells to be reprogrammed into iPSCs remain active. A possible solution would be to induce differentiation into the appropriate cell type prior to transplantation. Immune rejection of stem cells that originated from a source that is genetically distinct from the recipient would be a very common problem; a possible solution is to create iPSCs from the recipient's own cells. One source of heterogeneity in iPSC lines is the occurrence of random genetic mutations in the somatic cells of the donor or in the induced stem cells during their proliferation. This sort of heterogeneity can be uncovered by whole-genome sequencing but, more practically, by examining cellular morphology and rate of cell division by microscopic image analysis [3].

In spite of the great promise of stem cell therapy, only in a relatively few cases, its application has produced the expected results. The most successful cases include the use of hematopoietic stem cells for the treatment of blood-related malignancies and other disorders, and genetically modified stem cells for treatment of a few genetic diseases caused by single gene malfunction. In addition, skin stem cells (*keratinocytes*) have been used to treat severe burns and stem cells of the eye cornea to repair damage to this tissue. Skeletal disorders, including cartilage degeneration, osteoporosis and different forms of arthritis[1] are prime candidates for stem cell tissue repair. Therapies involve the site injection of *mesenchymal stem cells* (MSCs), multipotent cells found in a variety of embryonic areas (and in the umbilical cord) derived from the mesoderm embryonic germ layer [4,5]. The treatment of degenerative disorders of the nervous system, specifically of Parkinson's disease, has involved the use of cells from the

---

[1] *Osteoporosis* is the loss of bone density due to a greater activity of *osteoclasts* (the cells that degrade bone under normal conditions) than *osteoblasts* (the cells that synthesize bone matrix and effect its mineralization); *osteoarthritis* is the degeneration of the cartilage and the bone surfaces in skeletal joints due to the loss of function of *chondrocytes* (cells that synthesize cartilage); *rheumatoid arthritis* is an autoimmune disease that affects the joints and also other parts of the body.

mid-brain of embryos. Because the source of these cells is limited, pluripotent stem cells (PSCs) from the total cell mass of early embryos, cultured *in vitro* under conditions that ensure their differentiation into nerve cells, are now used. PSCs have also been tested in cases of spinal cord injuries and to repair the retinal damage caused by macular degeneration. An area of promise is the use of skeletal muscle stem cells to remedy the different types of muscular dystrophy and to enhance the process of muscle regeneration following injury. Muscle stem cells are present in the periphery of skeletal muscle fibers and use the same cascade of transcription factors for differentiation in the embryo as they do in the adult following injury. A major problem is that in most instances, these cells fail to become established in the damaged muscles and that muscles contain several cell types for which stem cells have not been identified. The different applications of stem cells for corrective and regenerative therapies just discussed will be perfected and greatly expanded by a thorough understanding of the basic genetic and epigenetic parameters responsible for stem cell differentiation and organismal development. For example, it is only recently that major epigenetic effects (the action of histone lysine demethylases) have been correlated with the differentiation of muscle stem cells [6].

In spite of these potential difficulties, some estimates predict that the regenerative medicine research and commercial enterprises will undergo an almost unlimited expansion.

Single-cell genetic analysis: The technical ability to determine the DNA sequence (genome) which genes are expressed (transcriptome) and the regulatory relationship between epigenetic marks and gene expression (epigenome) in single cells has revealed an unforeseen complexity in apparently uniform tissues. Single-cell "multiomics" analysis can provide the molecular blueprint and identify the many changes in gene expression used by cells to differentiate; it can be used to recreate this process experimentally as well as to predict the origins of transformation into cancer cells [7]. A major aspect of this exciting field in need of development concerns the accuracy of the data, for example, currently, in order to obtain the genome sequence and to be able to measure the transcription activity of single cells, the nucleic acids involved must be experimentally amplified, increasing substantially the probability of introducing errors. Another area in need of attention is the development of algorithms for the meaningful analysis of the gigantic mass of primary data generated.

Transgenerational inheritance: Another area of research is the transgenerational inheritance of epigenetic marks that have been induced by cellular or external environmental signals [8]. As previously mentioned (see Chapter 9), most of the evidence to date comes from experimental models ranging from roundworms (*Caenorhabditis*) to mice and rats. The environmental variables most commonly studied are diet, exposure to stress and exposure to toxins. To date, in non-human mammals, environmentally induced epigenetic changes do seem to persist over long time scales. Much less clear is the transgenerational inheritance of epigenetic characters in humans. Nevertheless, the influence of past environments on subsequent generations has such relevance to developmental biology, epidemiology and evolution theory that it should and will continue to be vigorously investigated.

Cancer immunotherapy: The interaction of cancer cells with the immune system is an area that will benefit from continued basic research into the molecular biology of cell differentiation and into the function of the complex network of cells and proteins that constitute the immune system. Cancer cells display molecular characteristics that distinguish them from the normal cells from which they arose and that could be used as tool for their immunological eradication. These new proteins are the result of mutations that occur only in cancer cells because of the latter's acquired genetic instability; for these reasons, they are considered foreign by the immune system and are ideal targets for immunotherapy in the form of tumor vaccines. Yet, cancers progress because of their cells' ability to avoid immunosuppression. This avoidance involves a variety of mechanisms including, for example, the epigenetic silencing in cancer cells of the genes responsible for the presentation of their novel characteristics to the immune system [9]. Another means of avoiding the immune system is to regulate the immune checkpoints that normally exist to prevent an over-response to pathological conditions, such as infections. Targeting these immune checkpoints is a current and promising area in cancer immunotherapy [10].

A highly promising area of research combines immunotherapy approaches with regenerative techniques. A recent breakthrough in this field is a cure for beta-thalassemia, a type of inherited blood disorder that causes a reduction of normal hemoglobin and red blood cells in the blood through mutations in the beta-globin gene. The treatment consists of isolating the patient's own bone marrow stem cells, genetically modifying them by introducing a functional beta-globin gene using a virus carrier and injecting them back into the patient [11].

References

[1] De Luca, M., *et al.*, Advances in stem cell research and therapeutic development. *Nature Cell Biology*, 2019. **21**(7): 801–811.

[2] Yamanaka, S., Pluripotent stem-cell based cell therapy — promise and challenges. *Cell Stem Cell*, 2020. **27**(4): 523–531.

[3] Hayashi, Y., K. Ohnuma, and M.K. Furue, Pluripotent stem cell heterogeneity. *Advances in Experimental and Medical Biology*, 2019. **1123**: 71–94.

[4] Kangari, P., *et al.*, Mesenchymal stem cells: Amazing remedies for bone and cartilage defects. *Stem Cell Research and Therapy*, 2020. **11**(1): 1–21.

[5] Humphreys, P.A., *et al.*, Developmental principles informing human pluripotent stem cell differentiation to cartilage and bone. *Seminars in Cell and Developmental Biology*, 2021. Article ID: 34949507, 20 pages.

[6] Cicciarello, D., L. Schaeffer, and I. Scionti, Epigenetic control of muscle stem cells: Focus on histone lysine demethylases. *Frontiers in Cell and Developmental Biology*, 2022. doi: 10.3389/fcell.2022.917771. 1–9.

[7] Wagner, D.E., and A.M. Klein, Lineage tracing meets single-cell omics: Opportunities and challenges. *Nature Reviews Genetics*, 2020. **21**(7): 410–427.

[8] Boskovic, A., and O.J. Rando, Transgenerational epigenetic inheritance. *Annual Review of Genetics*, 2018. **52**: 21–41.

[9] Cao, J., and Q. Yan, Cancer epigenetics, tumor immunity, and immunotherapy. *Trends in Cancer*, 2020. **6**(7): 580–592.

[10] Suraya, R., *et al.*, Immunotherapy in advanced non-small cell lung cancers: Current status and updates. *Cancer Management and Research*, 2022. **14**: 2079–2090.

[11] Locatelli, F., *et al.*, Betibeglogene autotemcel gene therapy for non–β^0/β^0 genotype β-Thalassemia. *The New England Journal of Medicine*, 2022. **386**: 415–427.

Index